AF361963

Symmetry in Complex Systems

Symmetry in Complex Systems

Editors

J. A. Tenreiro Machado
António M. Lopes

MDPI • Basel • Beijing • Wuhan • Barcelona • Belgrade • Manchester • Tokyo • Cluj • Tianjin

Editors
J. A. Tenreiro Machado
Institute of Engineering,
Department of Electrical Engineering,
Polytechnic Institute of Porto
Portugal

António M. Lopes
UISPA—LAETA/INEGI,
Faculty of Engineering,
University of Porto
Portugal

Editorial Office
MDPI
St. Alban-Anlage 66
4052 Basel, Switzerland

This is a reprint of articles from the Special Issue published online in the open access journal *Symmetry* (ISSN 2073-8994) (available at: https://www.mdpi.com/journal/symmetry/special_issues/Symmetry_Complex_Systems).

For citation purposes, cite each article independently as indicated on the article page online and as indicated below:

LastName, A.A.; LastName, B.B.; LastName, C.C. Article Title. *Journal Name* **Year**, *Article Number*, Page Range.

ISBN 978-3-03936-894-5 (Pbk)
ISBN 978-3-03936-895-2 (PDF)

Contents

About the Editors

J. A. Tenreiro Machado was born in 1957, graduated with 'Licenciatura' (1980), PhD (1989) and 'Habilitation' (1995) degrees in Electrical and Computer Engineering at the University of Porto. From 1980 to 1998, he worked as a professor at the Department of Electrical and Computer Engineering, University of Porto. Since 1998, he has been a coordinator professor at the Institute of Engineering of the Polytechnic Institute of Porto, Department of Electrical Engineering. His main research interests are: nonlinear dynamics, modeling, fractional calculus, evolutionary computing, control, and robotics. He is the author of more than 680 papers and his Scopus h-index is 54.

António M. Lopes obtained his PhD and 'Habilitation' in Mechanical Engineering in 2000 and 2018, respectively, both at the Faculty of Engineering of the University of Porto (FEUP), Portugal. Currently, he works at the Department of Mechanical Engineering (DEMec) and the Associated Laboratory for Energy, Transports and Aeronautics (LAETA/INEGI). António M. Lopes has been involved in teaching activities for the master's degree in Mechanical Engineering, the master's degree in Industrial Engineering and Management, and the doctoral program in Mechanical Engineering, all at FEUP. He participated in several research projects in the area of Mechanical Engineering and Automation. His research interests include: complex systems, fractional calculus, nonlinear dynamics, automation, robotics, control, systems modelling, and simulation. António M. Lopes has published about 130 papers in international journals with high impact factors, and nearly 100 book chapters, plus conference papers. His Scopus h-index is 23. He currently participates on the Editorial Board of eight international journals, and has been a guest editor of several journal Special Issues.

Editorial

Symmetry in Complex Systems

António M. Lopes [1,*] **and José A. Tenreiro Machado** [2]

1   UISPA–LAETA/INEGI, Faculty of Engineering, University of Porto, Rua Dr. Roberto Frias, 4200-465 Porto, Portugal
2   Institute of Engineering, Polytechnic of Porto, Department of Electrical Engineering, Rua Dr. António Bernardino de Almeida, 431, 4249-015 Porto, Portugal; jtm@isep.ipp.pt
*   Correspondence: aml@fe.up.pt

Received: 20 May 2020; Accepted: 3 June 2020; Published: 8 June 2020

Keywords: complex systems; symmetry-breaking; bifurcation theory; complex networks; nonlinear dynamical systems

---

Complex systems with symmetry arise in many fields, at various length scales, including financial markets, social, transportation, telecommunication and power grid networks, world and country economies, ecosystems, molecular dynamics, immunology, living organisms, computational systems, and celestial and continuum mechanics. The emergence of new order and structure in complex systems means symmetry breaking and transition from unstable to stable states. Modeling complexity attracted many researchers from different areas, dealing both with theoretical concepts and practical applications. This Special Issue seeks to fill the gap between the theory of symmetry-based dynamics and its application to model and analyze complex systems. This Special Issue focuses on the synergies between the theory of symmetry-based dynamics and its application to model and analyze complex systems. It includes 7 manuscripts addressing novel issues and specific topics that illustrate symmetry in complex systems. In the follow-up the selected manuscripts are presented in alphabetic order.

The manuscript "A Transmission Prediction Neighbor Mechanism Based on a Mixed Probability Model in an Opportunistic Complex Social Network" [1], by Genghua Yu, Zhigang Chen, Jia Wu and Jian Wu, proposes a routing decision method based on an improved probability model combined with a quantitative social relationship value and cooperative value to filter neighbor nodes. The algorithm combines multiple feature information between nodes and uses this feature information to quantify social relationship values and partnership values. Then, the prediction matrix is obtained by matrix decomposition and gradient descent, and the relay nodes are filtered according to the predicted probability values.

In the paper "Derivative Free Fourth Order Solvers of Equations with Applications in Applied Disciplines" [2], Ramandeep Behl, Ioannis K. Argyros, Fouad Othman Mallawi and J. A. Tenreiro Machado address efficient equation solvers for real- and complex-valued functions. The work extends earlier schemes and studies the computable radii of convergence and error bounds based on the Lipschitz constants. Furthermore, the range of starting points is explored to know how close the initial guess should be considered for assuring convergence.

In the work "Extending the Adapted PageRank Algorithm Centrality to Multiplex Networks with Data Using the PageRank Two-Layer Approach" [3], Taras Agryzkov, Manuel Curado, Francisco Pedroche, Leandro Tortosa and José F. Vicent propose a measure of centrality for biplex networks based on the adapted PageRank algorithm centrality for spatial networks with data. The scheme is implemented following the two-layers approach for PageRank model. The new measure of centrality can determine the importance of the nodes of a network and work with several data sets associated with the nodes themselves, without any connection or relationship between them.

The manuscript "Mei Symmetry and Invariants of Quasi-Fractional Dynamical Systems with Non-Standard Lagrangians" [4], by Yi Zhang and Xue-Ping Wang deals with quasi-fractional dynamical systems from exponential non-standard Lagrangians and power-law non-standard Lagrangians. Firstly, the definition, criterion, and corresponding new conserved quantity of Mei symmetry in this system are presented and studied. Secondly, considering that a small disturbance is applied on the system, the differential equations of the disturbed motion are established, the definition of Mei symmetry and corresponding criterion are given, and the new adiabatic invariants led by Mei symmetry are proposed and proved.

The paper "Multi-Agent Reinforcement Learning Using Linear Fuzzy Model Applied to Cooperative Mobile Robots" [5], by David Luviano-Cruz, Francesco Garcia-Luna, Luis Pérez-Domínguez and S. K. Gadi, presents a joint $Q-$function linearly fuzzified for a multi-agent system continuous state space, which overcomes the dimensionality problem. A proof for the convergence and existence of the solution proposed by the algorithm presented.

In the research "Time-Fractional Heat Conduction in a Plane with Two External Half-Infinite Line Slits under Heat Flux Loading" [6], Yuriy Povstenko and Tamara Kyrylych solve the time-fractional heat conduction equation with the Caputo derivative for an infinite plane with two external half-infinite slits with the prescribed heat flux across their surfaces. The integral transform technique is used. The solution is obtained in the form of integrals with the integrand being the Mittag-Leffler function.

The paper "Time-Fractional Heat Conduction in Two Joint Half-Planes" [7], by Yuriy Povstenko and Joanna Klekot, address the heat conduction equations with Caputo fractional derivative in two joint half-planes under the conditions of perfect thermal contact. The fundamental solution to the Cauchy problem as well as the fundamental solution to the source problem are examined. The Fourier and Laplace transforms are employed. The Fourier transforms are inverted analytically, whereas the Laplace transform is inverted numerically using the Gaver-Stehfest method.

The guest editors believe that the selected high-quality papers will help scholars and researchers to push forward the progress in the emerging area of symmetry in complex systems.

Funding: This research received no external funding.

Acknowledgments: The guest editors express their gratitude to the authors of the above contributions, and to the journal Symmetry and MDPI for their support during this work.

Conflicts of Interest: The authors declare no conflict of interest.

References

1. Yu, G.; Chen, Z.; Wu, J.; Wu, J. A Transmission Prediction Neighbor Mechanism based on a Mixed Probability Model in an Opportunistic Complex Social Network. *Symmetry* **2018**, *10*, 600. [CrossRef]
2. Behl, R.; Argyros, I.K.; Mallawi, F.O.; Tenreiro Machado, J. Derivative Free Fourth Order Solvers of Equations with Applications in Applied Disciplines. *Symmetry* **2019**, *11*, 586. [CrossRef]
3. Agryzkov, T.; Curado, M.; Pedroche, F.; Tortosa, L.; Vicent, J.F. Extending the Adapted PageRank Algorithm Centrality to Multiplex Networks with Data Using the PageRank Two-Layer Approach. *Symmetry* **2019**, *11*, 284. [CrossRef]
4. Zhang, Y.; Wang, X.P. Mei Symmetry and Invariants of Quasi-Fractional Dynamical Systems with Non-Standard Lagrangians. *Symmetry* **2019**, *11*, 1061. [CrossRef]
5. Luviano-Cruz, D.; Garcia-Luna, F.; Pérez-Domínguez, L.; Gadi, S.K. Multi-agent Reinforcement Learning using Linear Fuzzy Model Applied to Cooperative Mobile Robots. *Symmetry* **2018**, *10*, 461. [CrossRef]
6. Povstenko, Y.; Kyrylych, T. Time-fractional Heat Conduction in a Plane with Two External Half-infinite Line Slits under Heat Flux Loading. *Symmetry* **2019**, *11*, 689. [CrossRef]
7. Povstenko, Y.; Klekot, J. Time-Fractional Heat Conduction in Two Joint Half-Planes. *Symmetry* **2019**, *11*, 800. [CrossRef]

Article

A Transmission Prediction Neighbor Mechanism Based on a Mixed Probability Model in an Opportunistic Complex Social Network

Genghua Yu [1], Zhigang Chen [2,*], Jia Wu [2,*] and Jian Wu [1]

[1] School of Information Science and Engineering, Central South University, Changsha 410075, China; cnygh821@163.com (G.Y.); cnhnwujian@163.com (J.W.)
[2] School of Software, Central South University, Changsha 410075, China
* Correspondence: czg@mail.csu.edu.cn (Z.C.); jiawu5110@163.com (J.W.); Tel.: +86-133-8748-0797 (Z.C.); +86-182-7312-5752 (J.W.)

Received: 17 October 2018; Accepted: 1 November 2018; Published: 6 November 2018

Abstract: The amount of data has skyrocketed in Fifth-generation (5G) networks. How to select an appropriate node to transmit information is important when we analyze complex data in 5G communication. We could sophisticate decision-making methods for more convenient data transmission, and opportunistic complex social networks play an increasingly important role. Users can adopt it for information sharing and data transmission. However, the encountering of nodes in mobile opportunistic network is random. The latest probabilistic routing method may not consider the social and cooperative nature of nodes, and could not be well applied to the large data transmission problem of social networks. Thus, we quantify the social and cooperative relationships symmetrically between the mobile devices themselves and the nodes, and then propose a routing algorithm based on an improved probability model to predict the probability of encounters between nodes (PEBN). Since our algorithm comprehensively considers the social relationship and cooperation relationship between nodes, the prediction result of the target node can also be given without encountering information. The neighbor nodes with higher probability are filtered by the prediction result. In the experiment, we set the node's selfishness randomly. The simulation results show that compared with other state-of-art transmission models, our algorithm has significantly improved the message delivery rate, hop count, and overhead.

Keywords: Opportunistic complex social network; cooperative; neighbor node; probability model; social relationship

---

1. Introduction

With the mobile cellular network evolving to the fifth generation, huge amounts of information data are generated every day. Opportunistic complex social networks have become a common data delivery platform [1–3], and the popularity of mobile devices has enabled a variety of new services in social networks to be realized [4]. Many social tools, such as Google Plus [5], Facebook [6], and Twitter [7], have a large number of users and generate data in each moment. The traditional social network approaches to dealing with transmission and reception of big data diversification faces challenges due to the diversification of online data. Many wireless devices used to deliver information and then faced data overload, which would be a hindrance to information interworking and information sharing. In order to cope with the data transmission in 5G wireless networks, we need to design a model and more convenient transmission mode to implement data forwarding in a flexible manner, which is more suitable for complex network environments.

In opportunistic social networks, there is no complete end-to-end path between most mobile devices, so nodes communicate by hop by hop. One common method of data dissemination is a base station (BS) that can first deliver a message to a mobile device. The device of the user could carry the message and transmit it to other mobile devices via Device-to-Device(D2D) communication. The mobile device receives data and waits for an "opportunity" to deliver the message to the next mobile device [8]. For data forwarding problems, most existing work either assumes that users are completely unselfish, i.e., they are willing to help any user to pass information [9,10], or assume that users are absolutely selfish and that users need to be encouraged to participate in data transmission [11,12]. However, in real-life scenarios, users are not absolutely selfish or absolutely selfless. Their level of selfishness may be random or related to social relationships. Therefore, we have established a cooperative relationship to solve this problem.

However, the social network is a complex network. The challenge is it is hard to estimate the daily data sent and received by users [13,14]. It is difficult to satisfy a large amount of data decision, transmission, and storage by a personal PDA device alone. It may cause a low delivery ratio and an excessive energy consumption of the transmission device. The decision could be made by assistance devices, such as the base station or the edge. The delivery ratio can be improved and data transmission energy consumption can be reduced. In addition, this network is divided into multiple disconnected subnets due to the loss of nodes due to problems, such as device movement, failure, or non-cooperation [15,16]. Such nodes are called critical nodes. The loss of key nodes will cause the network to be disconnected, which leads to the loss of important data in the network [17,18]. The researcher proposed the concept of probabilistic routing [19]. Each node maintains a link probability table that reaches any other node in the network, and then it could determine the key nodes of message transmission by comparing probability sizes. However, this method is too simple to calculate probabilistic values in the case of an explosion of information. It is also too much trouble to perform calculations and decisions every time the data transmission node performs. Even the task of maintaining a routing table is too expensive. Therefore, we collected and used the node's encounter data, cooperation information, node characteristics, and social relationship to assist decision-making to predict key nodes in the opportunistic social network, and propose a relatively simple transmission model based on probability prediction.

To solve the huge amount of information in the 5G network, the mobile device has too much energy consumption for decision-making, transmission, and storage of a large amount of data. We propose ways to assist decision-making information transmission by devices, such as base stations or edges. Since the opportunistic complex social network does not necessarily acquire information about all nodes in the entire network, the node information is incomplete and the characteristics of each node cannot be utilized more fully to filter the key nodes. Therefore, the collected network information may not be complete, and the resulting encounter probability matrix may be incomplete or even sparse. In response to this problem, we first calculated some node social relationship values, cooperation probability values, and node encounter probabilities by pre-processing some data collected over a period of time. Then, we establish three related feature matrices and the vector of residual energy of nodes, and propose an algorithm based on hybrid probability matrix decomposition to predict key nodes. Finally, we filter the key nodes more reasonably and effectively through the impact of the collected feature information on the message passing between nodes in the network. Specifically, hidden information between nodes can also be fully utilized through the model to maximize the value of the node. By adaptive message forwarding decisions, the purpose of selecting a relay node and improving network resource utilization is achieved.

Specifically, the main contribution of this paper can be summarized as the following three aspects:

1. Through the transmission of decision information assisted by base stations or edge devices, we propose a method of pre-processing collected device information in an opportunistic complex social network and calculating social relationship values and cooperation probability values, so that the new probability matrix can be trained by adding additional information.
2. We propose a hybrid probability matrix decomposition model to predict the probability of encountering a node. We add node encounter information, social relationships, and partnerships

to form a hybrid model to predict the encounter cooperation (EC) values between nodes, and filter the key encounter nodes through EC values.

3. We have designed a simpler way to transfer information to share the large number of transmission tasks of the central equipment. The mobile device only needs to request the central node to encounter the probability table of the other node and the destination node. Then, according to the information in the table, the data is passed to the neighbor node that has a higher value than its own and the destination node. The simulation results show that the PNEC algorithm has excellent effects in the message transmission process between nodes, and maximizes the characteristics of mobile nodes in the network.

2. Related Work

In the form of 5G networks, base station construction is relatively dense. The base station needs to collect the soaring data generated by devices, and analyze, process and transmit the data further. As shown in [20], a practical system architecture is proposed to process and extract useful information, and through the data caching mechanism, to solve the user data requirements. This approach increases the complexity of data processing. In such cases, most of the work of data transmission and data sharing can be done by the mobile device. The node's transmission decision, calculation, and other tasks can be assigned to the base station or edge equipment to reduce the node energy consumption. Many researchers have proposed ways for mobile devices to deliver messages in opportunistic mobile complex social networks. In such a network, there is no guarantee that the full connection path between the source and destination exists at any time, which makes the traditional routing protocol unable to deliver messages better between nodes. Some researchers have proposed probabilistic routing protocols for the network [19]. The nodes predict the reachability probability between the nodes by storing their encounter information, and select the nodes with higher encounter probability as the relay node. Such a method will increase the energy consumption of nodes in the case of the heavy task of 5G network data transmission.

In mobile social networks, some researchers have proposed using community structure to make data forwarding decisions. They assume that people in the same community would have closer relationships and more opportunities to connect with each other. The concept of the probability of the source node to the destination node is proposed by tracking the encounter situations of the nodes to study the activity of the node and the probability of reaching the target node, using the Poisson model of the social network contact. On this basis, the contact-aware opportunity-based forwarding (CAOF) scheme [21] is proposed by calculating the global and local probability of two different transmission phases. It includes the forwarding scheme between the inter-community transmission phase and the intra-community transmission phase. In the inter-community phase, neighbor nodes with higher global activity and high probability of source node to target node are selected as relays. Furthermore, forwarding decisions are determined by local metrics in the internal phase of the community. In this program, social characteristics and cooperation relationships may not be considered for integration into the design of the scheme, and may not perform well when the nodes are selfish.

In addition, there is an interesting study on mobile social networks. Since the mobile social network consists of nodes with different social attributes, the connection is transmitted or shared by chance. Some researchers have proposed a routing algorithm based on social identity awareness (SIaOR) [22]. They believe that many socially aware routing algorithms ignore an important social attribute, social identity. Therefore, researchers propose an opportunity routing algorithm in mobile social networks by considering the social identity and social impact of mobile nodes in mobile social networks. The algorithm not only considers the multiple social identities of mobile nodes, but also their social impacts. However, it is difficult to measure the ability of nodes to forward data if they simply consider the social identity impact with the target node. Moreover, nodes with strong social relationships may not be suitable nodes for cooperation.

A new sensing method is the Internet of Things in the Mobile Crowd Sensor Network (MCSN). There are two existing transmission mechanisms: One is to transmit data through a cellular network, the other is an opportunistic transmission method through short-range wireless communication technology. Some researchers have proposed that the cost of cellular network transmission is high and it is not conducive to increasing user participation. Therefore, researchers focus on the application of wireless communication technology in MCSN by constructing a new opportunistic propagation model. They proposed an opportunistic data transmission mechanism based on the Socialization Susceptible Infected Susceptible Epidemic Model (SSIS) [23]. The mechanism uses the SSIS model to obtain a social relationship table by analyzing the social relationships of mobile nodes. The source node that performs the sensing task through the combination of the social relationship table and the spray and wait mechanism selectively propagates the data to other nodes until it reaches the destination node that can send the data to the platform. Using the spray and wait method will make the transmission mode more complicated and the node energy consumption increase.

Some researchers have proposed a new mobile opportunistic network routing protocol, MLProph [24], through machine learning and through further research on the mobile social network. The model uses decision trees and neural networks to train various factors, such as the predicted value inherited from the PROPHET routing scheme, node popularity, node's power consumption, speed, and location, to further calculate the probability of successful delivery of information. The algorithm trains based on historical information to obtain an equation for calculating the probability to detect whether the node can send the message to the destination node. This is an interesting way to make probabilistic predictions through machine learning methods, however, this method may only consider the nodes themselves, without considering the characteristics between the nodes.

According to the discussion on those methods, there is no complicated decision-making scheme considering the cooperative relationship and social relationship of the nodes. Additionally, the key neighbors are calculated and decided by nodes. When the amount of data transmission is large, calculation and decision by nodes may increase energy consumption. We need more complicated decision-making methods to solve data transmission problems more conveniently when the amount of data skyrockets in 5G networks. A good and effective decision-making mode determines whether the information transmission between mobile devices can achieve the desired effect. Many researchers have proposed a probabilistic routing method to calculate the probability of a nodes' final successful delivery to the destination by training the probabilistic equation. However, these methods could not consider whether the selected relay node is willing to cooperate in forwarding messages. Furthermore, nodes move randomly and socially, and it is worth considering whether such social relations will affect the activity rules of nodes. Therefore, we designed a model based on probabilistic prediction by quantifying social relationships and cooperative relationship values. We need to analyze and process and make decisions on a large amount of data, and improve the user satisfaction through active caching of edge devices to meet the needs of users in the 5G network [20]. In this case, we can make transmission calculations and decisions through small base stations or edge devices to reduce the workload of mobile nodes. Consequently, we have collected the characteristic information of nodes in the region through the base station and quantified the collected information to form the encounter matrix, cooperative matrix, and social relation matrix, respectively. Then, we used the improved probability matrix decomposition method to integrate the matrix information to update the encounter probability matrix and predict the probability value of the encounter and cooperation between two nodes without the encounter history information and updated the existing value. Finally, the neighbor node is filtered by the size of EC in the updated probability matrix. Experiments show that the proposed model reduces the network overhead of non-cooperative nodes, optimizes the path of messages to the destination nodes, and enables messages to be transmitted along the cooperative nodes with a high probability of encountering the destination nodes.

3. Model Design

3.1. Node Data Collection and Transmission

The base station (BS) collects information of all nodes in the area and trains the encounter matrix at period, T, in the area. As shown in Figure 1, the base station collects the information of all nodes in the region over a period of time and trains the final probability matrix according to the designed model through these features. If the social relationship and the cooperative relationship cannot be obtained in the initial stage of the model application, the final probability matrix is the historical encounter probability matrix. When a node has a transmission task, the probability table of known nodes and destination nodes is requested from the base station. Specifically, when a new node enters the area, its characteristic information is sent to the BS. If it carries a probability forecast table, the BS obtains the table and updates its own matrix based on the probability forecast table obtained from the BS. If it does not carry a probability forecast table, it is sent to the matrix trained by the BS in the area. After the T period, the BS trains the matrix, M, according to Algorithm 1 (refer to Section 3.5), and transmits the matrix information according to the node information collected in the past T periods. During the period, the matrix, M, is updated by Algorithm 2 (refer to Section 3.5) according to the obtained node's probability forecast table and communication between BS before the T period. The main update idea is to add a record without some node information. The exchange of information when the mobile devices meet is also the same as updating the records carried. The message carried by the mobile devices is sent to the encounter devices whose probability of encounter is greater than the mobile device's own and destination mobile device's probability of encounter. If there is no record of the destination node, it is sent to nodes in the encounter node that have a higher probability of encountering mobile devices in another area.

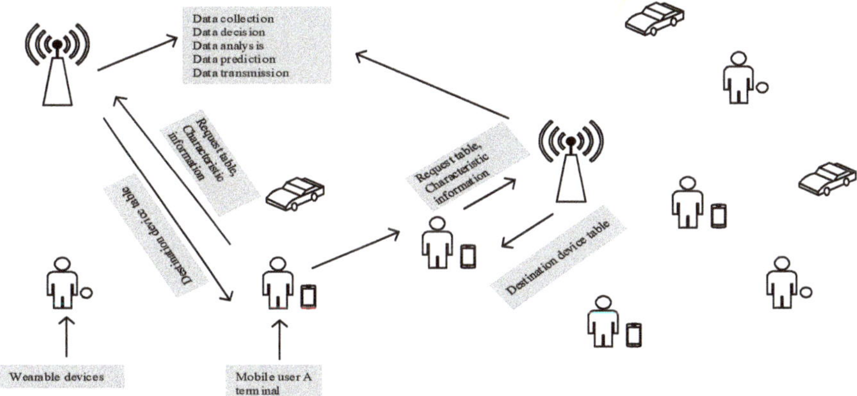

Figure 1. Example diagram of node information exchange in the area.

3.2. Encounter Probability and Social Relationship Decomposition

3.2.1. Encounter Relationship Value Calculation

First, we give a definition of computing the encounter probability. To facilitate the description of related problems, in this paper, m_{ij} is the probability of an encounter between node i and node j over a period of time. That is, the probability that node i and node j meet within the perceived range. The definition is as follows:

$$m_{ij} = \frac{w_{ij}}{\sum\limits_{adj}^{n} w_{i,adj}} \tag{1}$$

where i denotes the source node, and j denotes the target node. w_{ij} is the number of historical encounters between node i and node j within a certain period of time. $w_{i,adj}$ indicates that the number of times that node i has met with other nodes is within a certain period of time.

3.2.2. Social Relationship Value Calculation

The number of encounters between nodes simply reflects the node's encounters over a period of time. We cannot just rely on the number of node history encounters to predict an encounter in the future. Here, we consider the quantitative factors of mobile node social relations.

(1). Node Similarity

We believe that the closer the social relationship is, the more likely it is to meet in the future. We first consider the number of shared neighbor nodes in the node's social relationship. The similarity between nodes is defined as the number of common neighbors between two nodes. This equation can be defined as:

$$sim_{ij} = \frac{|C_i \cap C_j|}{|C_i \cup C_j|} \tag{2}$$

where sim_{ij} represents the similarity between node i and node j. $|\cdots|$ represents the number of nodes in the collection. C_i is the set of neighbor nodes connected to node i at the current time, and C_j is the set of neighbor nodes connected to node j at the current time. Considering that two nodes share a similar set of neighbor nodes, two nodes are more likely to pass data through a common neighbor. Therefore, we consider that the relationship between the two nodes is higher and the probability of data transmission is higher.

(2). Devices' Mobility

Considering other factors that affect the value of total social relationships, node mobility change and node connection transformation are important factors that must be considered in the value of the overall social relationship. In a mobile opportunistic network, since the nodes are constantly moving, the connection between the nodes and the neighbors will change over time. In a continuous T time interval, the degree of change in the movement of such a node relative to another node is defined as the degree of motion transformation:

$$Move_{ij} = \frac{\left|(C_i' \cup C_j') \cup (C_i \cup C_j)\right| - \left|(C_i' \cup C_j') \cap (C_i \cup C_j)\right|}{\left|(C_i' \cup C_j') \cup (C_i \cup C_j)\right|} \tag{3}$$

where $Move_{ij}$ is the moving degree of the node at the current time. C_i' represents the set of neighbor nodes of node i before the T time interval, as is the case with C_j'. C_i represents the set of neighbor nodes of node i at the current time, and the same is true for C_j. From Equation (3), we can deduce that the higher the frequency of the movement of node j relative to node i, the larger the transformation of its neighbor node set relative to i, and the greater the degree of motion transformation.

(3). Connection Transformation

The degree to which a shared neighbor node of a node's connection with respect to another node changes and the dynamics of the neighbor node connection between them are referred to as the connection degrees of transition. In successive T time intervals, the degree of change in the connection of such a node relative to another node is defined as the connection transformation:

$$Conn_{ij} = \frac{\left|(C_i' \cap C_j') \cup (C_i \cap C_j)\right| - \left|(C_i' \cap C_j') \cap (C_i \cap C_j)\right|}{\left|(C_i' \cap C_j') \cup (C_i \cap C_j)\right|} \tag{4}$$

From the formula (4), $Conn_{ij}$ is the degree of change in the shared neighbor node set connected to the node. It can be seen that the greater the change in the set of shared neighbor nodes of the connection is, the higher the value of $Conn_{ij}$. By observing and analyzing the common neighbor nodes of the two nodes, we can see the motion changes of the two nodes in the current network. When the common neighbor node with node i associated with node j increases, the connection probability between node i and node j will increase, and then the possibility of data exchange will be improved.

Through the analysis of node similarity and node relative change, we can quantify the social relationship values between two nodes as follows:

$$S_{ij} = \alpha sim_{ij} + \beta(1 - Move_{ij}) + \gamma(1 - Conn_{ij}) \tag{5}$$

Among them, the smaller the degree of transformation, the greater the value of social relations. α, β, γ are the coefficients of node similarity, moving transform, and connection transform, respectively. They represent the weighting factors that influence the value of social relationships by different factors, and $\alpha + \beta + \gamma = 1$.

3.2.3. Decomposition Method

We represent the probability of encounter and social relationship values between the nodes described above in a matrix. Wherein, let $M = \{m_{ij}\}$ denote the matrix of $n \times n$; that is, the encounter probability matrix. For a pair of nodes, n_i and n_j, $m_{ij} \in [0,1]$ denote the historical encounter probability of node i to node j. Let $S = \{s_{ij}\}$ denote the social relationship matrix of $n \times n$. Two nodes with strong social relationships affect each other's probability of encountering the same node. We also believe that nodes are more willing to be close to the nodes with high social relations. Because the devices in the opportunistic network are mostly carried by people, it is significant to analyze the probability of node encounters through the social relations of nodes. It is assumed that node, n_a, knows nothing about a node, n_c, and it meets node, n_b, and node, n_c, with a high relation value of node, n_a. So, node, n_a, meets it because it is close to node, n_b, and then node, n_a, and node, n_c, also meet. From the above analysis, we can summarize the above social process as:

$$\widetilde{M}_{ij} = \frac{\sum\limits_{z \in \kappa(i)} M_{zj} S_{iz}}{|\kappa(i)|} \tag{6}$$

where $\widetilde{M}_{ij}$ is the prediction of the probability that user u_i meets user u_j, M_{ij} is the probability that user u_i meets user u_j, $\kappa(i)$ is the neighbor's set that user u_i relations and $|\kappa(i)|$ is the number of related users of user u_i in the set, $\kappa(i)$. $|\kappa(i)|$ can be merged into S_{ij}, since it is the normalization term of the relation evaluation of the estimate.

Then, we can infer the prediction of the probability that user u_i for all users is as follows:

$$\begin{pmatrix} \widetilde{M}_{i1} \\ \widetilde{M}_{i2} \\ \vdots \\ \widetilde{M}_{in} \end{pmatrix} = \begin{pmatrix} S_{i1} & S_{i2} & \cdots & S_{in} \end{pmatrix} \begin{pmatrix} M_{11} & M_{12} & \cdots & M_{1n} \\ M_{21} & M_{22} & \cdots & M_{2n} \\ \vdots & \vdots & \ddots & \vdots \\ M_{n1} & M_{n2} & \cdots & M_{nn} \end{pmatrix} \tag{7}$$

Consequently, for all users, we can infer as:

$$\widetilde{M} = M^T S \tag{8}$$

where SM can be interpreted as the probability forecasting purely based on the user relevance.

The idea of the probability matrix decomposition is to derive the l-dimensional features of the high-quality representative node (or user), N, of the node (or user) based on the analysis of the probability values of the encounter between the nodes. Let $N \in M^{l \times n}$ and $X \in M^{l \times n}$ be the latent nodes and social factors. The feature matrix, the column vectors, N_i and X_j, represent the potential feature vectors of the node and social factor, respectively. We define the conditional distribution of the observed encounter probability as:

$$p(M|N, X, S, \sigma_M^2) = \prod_{i=1}^{n} \prod_{j=1}^{n} \mathcal{N} \left[(m_{ij}|g(\sum_{z \in \kappa(i)} S_{iz} N_z^T X_j), \sigma_M^2) \right]^{I_{ij}^M} \tag{9}$$

where $\mathcal{N}(x|u, \sigma^2)$ is the probability density function of the Gaussian distribution with a mean of u and a variance of σ^2, and I_{ij}^M is an exponential function. If node i meets node j, it is equal to 1, otherwise it is equal to 0. The function, $g(x) = 1/(1 + \exp(-x))$, is a logarithmic function, which allows the function value corresponding to $N_i^T X_j$ to fall in the interval $[0, 1]$. We use zero-mean spherical Gaussian priors [24,25] on the node and social factor feature vectors:

$$p(N, \sigma_N^2) = \prod_i^n \mathcal{N}(N_i|0, \sigma_N^2 I) p(X, \sigma_X^2) = \prod_j^n \mathcal{N}(X_i|0, \sigma_X^2 I) \tag{10}$$

So, through the node and social factors features with a simple Bayesian inference [26] in Formulas (9) and (10), we have:

$$\begin{aligned} &p(N, X|M, S, \sigma_M^2, \sigma_N^2, \sigma_X^2) \\ &\propto p(M|S, N, X, \sigma_M^2) p(N|\sigma_N^2) p(X|\sigma_X^2) \\ &= \prod_{i=1}^{n} \prod_{j=1}^{n} \mathcal{N}[(m_{ij}|g(\sum_{z \in \kappa(i)} S_{iz} N_z^T X_j), \sigma_M^2]^{I_{ij}^M} \\ &\times \prod_{i=1}^{n} \mathcal{N}(N_i|0, \sigma_N^2 I) \times \prod_{j=1}^{n} \mathcal{N}(X_j|0, \sigma_X^2 I) \end{aligned} \tag{11}$$

We can assume that S is independent with the low-dimensional matrices, N and X, then this Equation (11) can be changed to:

$$\begin{aligned} &p(N, X|M, S, \sigma_M^2, \sigma_N^2, \sigma_X^2) \\ &= \prod_{i=1}^{n} \prod_{j=1}^{n} \mathcal{N}[(m_{ij}|g(\lambda N_i^T X_j + (1 - \lambda) \sum_{z \in \kappa(i)} S_{iz} N_z^T X_j), \sigma_M^2]^{I_{ij}^M} \\ &\times \prod_{i=1}^{n} \mathcal{N}(N_i|0, \sigma_N^2 I) \times \prod_{j=1}^{n} \mathcal{N}(X_j|0, \sigma_X^2 I) \end{aligned} \tag{12}$$

3.3. Cooperation Probability and Energy Decomposition

3.3.1. Cooperative Probability Calculation

(1). Node Preference

$Pref_{ij}$ is the degree of preference between the two nodes, i and j. The node preference is different from the node encounter frequency, which refers to the time occupation ratio between two nodes establishing a connection and transmitting a message within a period of time. The specific calculation formula is as follows:

$$Pref_{ij} = \frac{\sum_{k=1}^{n} t_{i,j}^k}{\sum_{k=1}^{n} t_{i,other}^k} \tag{13}$$

$t_{i,j}^k$ represents the total time taken for message delivery when node i and node j establish a connection for the kth time. $t_{i,other}^k$ represents the total time taken for message delivery when node i and other nodes establish a connection for the kth time.

(2). Node Relationship Stability

Rel_{ij} is the correlation between the average interval time delivery messages of the two nodes, i and j. Node relationship stability refers to the average interval between messages sent between nodes. Through it we can predict the possibility of future nodes establishing connections for a long period of time. This means that the shorter the connection interval in the most recent period, the more likely it is that the connection will be established in the future. In a period of time, the more the number of encounters, the longer the interval. Which indicates that the node has stable cooperation for a long time, as described below:

$$Rel_{ij} = \frac{T_{i,j}^{k} - T_{i,j}^{1}}{k \times T} \tag{14}$$

In the above equation, k is the number of message passes in T time. $T_{i,j}^{k}$ and $T_{i,j}^{1}$ are the time of the kth transmission of the message and time of the first transmission of the message, respectively, in T time.

(3). Node Activeness

We use the number of times the node forwards the message for a period of time as a measure of node activeness. The higher the node activity, the higher the degree of cooperation is to some extent. We define it as follows:

$$Act_i = \frac{re_i}{\sum\limits_{i=1}^{n} re_i} \tag{15}$$

where re_i indicates that the number of times the node i forwards the message is within a certain period of time, T.

(4). Data Demand

We consider that nodes are more willing to cooperate to deliver data with common needs. At the time of message delivery, the node gives a rating feedback of the message demand. Whether a node meets and connects with other nodes within a certain period of time, the node prefers to receive and forward data useful to itself. Therefore, if the score of node feedback is higher for a period of time, it indicates that the probability that the neighbor node is willing to cooperate to forward the message is larger. Increasing this parameter can help predict the likelihood of future cooperation between nodes.

$$Need_{ij} = \frac{\sum\limits_{\tau=1}^{k} q_{ij}^{\tau}}{Q \times k} \tag{16}$$

where q_{ij}^{τ} represents the data demand feedback score given by node j when node i passes a message to node j. Q represents the highest score.

Through the degree of node preference, relationship stability, activity, and data demand analysis, we can quantify the cooperation probability values between two nodes as follows:

$$R_{ij} = aPref_{ij} + bRel_{ij} + cAct_j + dNeed_{ij} \tag{17}$$

Among them, a, b, c, d is the coefficient of node preference, relationship stability, activity, and data demand. They respectively represent the weighting factors that influence the value of cooperation and $a + b + c + d = 1$.

3.3.2. Residual Energy Ratio

We consider the residual energy of the current moment of the mobile node. When the remaining energy of the mobile node is greater, the node is more willing to cooperate to forward the data.

$$E_i = \frac{E_r}{E_d} \tag{18}$$

where E_r represents the energy remaining in the device and E_d is the energy required for the packet to be transmitted.

3.3.3. Cooperation and Energy Decomposition

We consider the cooperation probability matrix. Suppose we have n nodes. The probability of forwarding messages from n nodes to different nodes is different. In other words, the selfishness of nodes for different nodes is different. We use the above method to calculate the cooperation value to form a cooperative forwarding probability matrix. Considering that the cooperation relationship between nodes is related to the ratio of the remaining energy of nodes, we believe that the greater the residual energy ratio of nodes, the more stable the cooperative relationship of nodes. It is significant to analyze the probability of node cooperation through the residual energy ratio of nodes. From the above analysis, we can define this as:

$$\tilde{R} = E^T \circ R \tag{19}$$

where E is a vector with a residual energy ratio of n nodes, and the $\circ$ symbol is defined as each element corresponding to the i-th column element multiplied by the ith column of R. The new matrix we get is still represented by R, which is $R = \tilde{R}$.

Therefore, we let r_{ik} represent the probability of the message node, n_k, forwarding sent by node, n_i. We set $N \in R^{l \times n}$ and $Y \in R^{l \times n}$ as the feature matrix of potential nodes and cooperation factors, and the column vectors, N_i and Y_k, respectively represent potential feature vectors of the node and cooperation factor. We define the conditional distribution of the observed cooperative probabilities as:

$$p(R|N,Y,\sigma_R^2) = \prod_{i=1}^{n}\prod_{k=1}^{n} \mathcal{N}\left[(r_{ik}|g(\sum_{z \in \kappa(i)} S_{iz}N_z^T X_{jk}), \sigma_R^2) \right]^{I_{ij}^R} \tag{20}$$

where I_{ij}^R is an exponential function. If node n_i has the probability of forwarding the message sent by node n_k, it is equal to 1, otherwise it is equal to 0. We also place zero-mean spherical Gaussian priors [24,25] on the node and cooperation factor feature vectors:

$$p(N,\sigma_N^2) = \prod_{i}^{n} \mathcal{N}(N_i|0,\sigma_N^2 I) \, p(Y,\sigma_Y^2) = \prod_{k}^{n} \mathcal{N}(Y_k|0,\sigma_Y^2 I) \tag{21}$$

Therefore, through the node and cooperation factors features with a simple Bayesian inference [26] in Formulas (20) and (21), we have:

$$\begin{aligned}
&p(N,Y|R,\sigma_R^2,\sigma_N^2,\sigma_Y^2) \\
&\propto p(R|N,Y,\sigma_R^2)p(N|\sigma_N^2)p(Y|\sigma_Y^2) \\
&= \prod_{i=1}^{n}\prod_{k=1}^{n} \mathcal{N}\left[(r_{ik}|g(N_i^T Y_j),\sigma_R^2) \right]^{I_{ik}^R} \times \prod_{i=1}^{n} \mathcal{N}(N_i|0,\sigma_N^2 I) \times \prod_{k=1}^{n} \mathcal{N}(Y_k|0,\sigma_Y^2 I)
\end{aligned} \tag{22}$$

3.4. Hybrid Model Solution

Through the above processes, to reflect the probability that the historical encounter between nodes will affect the judgment that the node is more likely to meet a certain node, we use the graphical model shown in Figure 2 to model the problem of probability prediction. The model combines the encounter probability matrix, social relationship matrix, and node cooperative probability matrix into a consistent and compact feature representation. According to Figure 2, and Equations (11) and (22), the logarithm of the posterior distribution of node encounters and cooperation probability matrix are given by:

$$
\begin{aligned}
\ln p(N, X, Y | M, R, \sigma_M^2, \sigma_R^2, \sigma_N^2, \sigma_X^2, \sigma_Y^2) = \\
-\frac{1}{2\sigma_M^2} \sum_{i=1}^{n} \sum_{j=1}^{n} I_{ij}^M \left(m_{ij}^* - g(\lambda N_i^T X_j + (1-\lambda) \sum_{z \in \kappa(i)} S_{iz} N_z^T X_j) \right)^2 \\
-\frac{1}{2\sigma_R^2} \sum_{i=1}^{n} \sum_{k=1}^{n} I_{ik}^R (R_{ik} - g(N_i^T Y_k))^2 - \frac{1}{2\sigma_N^2} \sum_{i=1}^{n} N_i^T N_i - \frac{1}{2\sigma_X^2} \sum_{j=1}^{n} X_j^T X_j - \frac{1}{2\sigma_Y^2} \sum_{k=1}^{n} Y_k^T Y_k \\
-\frac{1}{2} \left(\left(\sum_{i=1}^{n} \sum_{j=1}^{n} I_{ij}^M \right) \ln \sigma_M^2 + \left(\sum_{i=1}^{n} \sum_{k=1}^{n} I_{ik}^R \right) \ln \sigma_R^2 \right) - \frac{1}{2} \left(nl \ln \sigma_N^2 + nl \ln \sigma_X^2 + nl \ln \sigma_Y^2 \right) + \varepsilon
\end{aligned}
\tag{23}
$$

where ε is a constant that does not depend on parameters. The hyperparameters (i.e., observed noise variance and a priori variance) are kept fixed after maximizing the logarithmic values of the three potential features, which is equivalent to the use of quadratic regularization terms to minimize the following sum of squared error objective functions:

$$
\begin{aligned}
L(R, M, N, X, Y, S, T) \\
= \frac{1}{2} \sum_{i=1}^{n} \sum_{j=1}^{n} I_{ij}^M \left(m_{ij} - g(\lambda N_i^T X_j + (1-\lambda) \sum_{z \in \kappa(i)} S_{iz} N_z^T X_j) \right)^2 \\
+ \frac{\lambda_R}{2} \sum_{i=1}^{n} \sum_{k=1}^{n} I_{ik}^R (r_{ik} - g(N_i^T Y_k))^2 + \frac{\lambda_N}{2} \|N\|_F^2 + \frac{\lambda_X}{2} \|X\|_F^2 + \frac{\lambda_Y}{2} \|Y\|_F^2
\end{aligned}
\tag{24}
$$

About formula (24), $\lambda_R = \sigma_M^2 / \sigma_R^2$, $\lambda_N = \sigma_M^2 / \sigma_N^2$, $\lambda_X = \sigma_M^2 / \sigma_X^2$, $\lambda_Y = \sigma_M^2 / \sigma_Y^2$, $\| \cdot \|_F^2$ is the F norm. The local minimum of the objective function given by Equation (24) can be found by performing gradient descent in N_i, X_j, and Y_k.

$$
\begin{aligned}
\frac{\partial L}{\partial N_i} = \; & \alpha \sum_{j=1}^{n} I_{ij}^M g'(\alpha N_i^T X_j + (1-\alpha) \sum_{z \in \kappa(i)} S_{iz} N_z^T X_j) X_j \\
& \times (g(\alpha N_i^T X_j + (1-\alpha) \sum_{z \in \kappa(i)} S_{iz} N_z^T X_j) - m_{ij}) \\
& + (1-\alpha) \sum_{h \in \varphi(i)} \sum_{j=1}^{n} I_{hj}^M g'(\alpha N_h^T X_j + (1-\alpha) \sum_{z \in \kappa(h)} S_{hz} N_z^T X_j) \\
& \times (g(\alpha N_h^T X_j + (1-\alpha) \sum_{z \in \kappa(h)} S_{hz} N_z^T X_j) - m_{hz}) S_{hi} X_j \\
& + \lambda_R \sum_{k=1}^{n} I_{ik}^R g'(N_i^T Y_k)(g(N_i^T Y_k) - r_{ik}) Y_k + \lambda_N N_i \\
\frac{\partial L}{\partial X_j} = \; & \sum_{i=1}^{n} I_{ij}^M g'(\lambda N_i^T X_j + (1-\lambda) \sum_{z \in \kappa(i)} S_{iz} N_z^T X_j) \\
& \times \left(g(\lambda N_i^T X_j + (1-\lambda) \sum_{z \in \kappa(i)} S_{iz} N_z^T X_j) - m_{ij}^* \right) \times \left(\lambda N_i + (1-\lambda) \sum_{z \in \kappa(i)} S_{iz} N_z^T \right) + \lambda_X X_j \\
\frac{\partial L}{\partial Y_k} = \; & \lambda_R \sum_{i=1}^{n} I_{ik}^R g'(N_i^T Y_k)(g(N_i^T Y_k) - r_{ik}) N_i + \lambda_Y Y_k
\end{aligned}
\tag{25}
$$

where $\varphi(i)$ is the set of trust nodes for node n, and $g'(x)$ is the derivative of the logarithmic function, $g'(x) = \exp(-x)/(1 + \exp(-x))^2$.

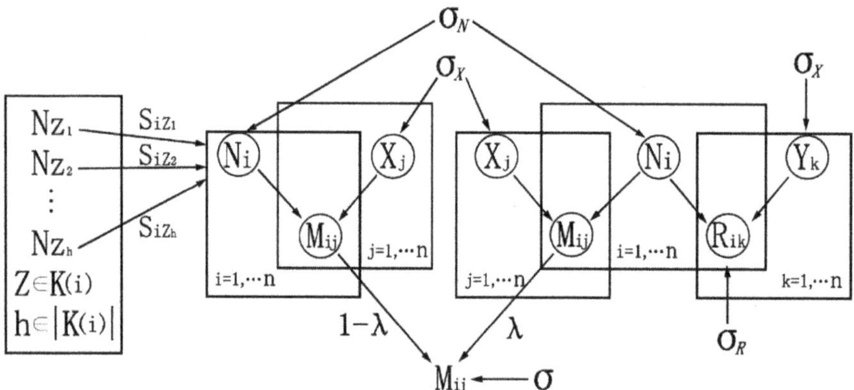

Figure 2. Probability prediction model.

3.5. Algorithm

The new probability matrix is obtained by using the matrix of social relation transformation and the matrix of cooperation forwarding relation. Because meeting the probability matrix is obtained by node meet history information, which is often incomplete, we need to fill in the missing value with additional information or update the encounter value with the cooperation degree and social of the devices. We called the newly updated probability matrix the encounter cooperation matrix containing the encounter cooperation (EC) values. Our algorithm is shown as follows:

Algorithm 1: PEBN (predict the probability of encounters between nodes).

Input: M: The encounter matrix formed by the node encounter probability map in the T period
$Sim, Move, Conn, Pref, Rel, Act, Need, E$

       T period

  1:   **Let** $t = 0$; //t is used to record the elapsed time after the new matrix is updated;
         Let A [ ]; //Record all nodes that carry the message.

  2:   $S \leftarrow getSocialMatrix(Sim, Move, Conn, T)$ //get the social relations matrix

  3:   $R \leftarrow getCooperationMatrix(Pref, Rel, Act, Need, E, T)$ //get the cooperation relations matrix

  4:   $\tilde{M} \leftarrow getMeetProbablityMatrix(M, S, R, T)$ //The new probability matrix of meeting is calculated
         based on the historical meeting, social relations and cooperation relations of the past T time period

  5:   $M^T \leftarrow getUpdateMatrixAccordAlgorithm2(M)$

  6:   **For each** $i \in (1, A.length]$ **do**

  7:   Create a set of neighbor nodes N_{adj} //Devices in A looks for the neighbor within the communication
         range after Δt time.

  8:   **If** N_{adj} contains a destination node; break; //If neighbors contains a destination node, the
         information transmission is successful and the loop ends.

  9:   **Else If** M^T has the column vectors of the destination node **Then**

10:   **For each** $z \in \left[1, N_{adj}.length\right]$

11:      **If** the EC value of the node n_z in N_{adj} and destination node> the EC value of node n_i and
         destination node //The neighbor with the highest EC Value of the target node is selected as the next
         hop node.

12:         Choose next hop nodes n_z

13:         A. add (n_z); //The selected devices are used as a new message-carrying devices.

14:      **End if**

15:     **End for**

16:    **End if**

17:   **Else If** M^T does not have the column vectors of the destination node **Then**

18:   $M' \leftarrow M^T.getRemoveColVector(M)$ //The nodes contained in M are the area. We need to find
         nodes with high probability of encountering other area and remove the column vectors
         corresponding to the area nodes.

19:   $N[] \leftarrow N_{adj}.getChooseColMaxMR(M', M'.getLine(N_{adj}, A.get(i))$ //If the corresponding EC value
         of node n_i is the largest, this column is invalidated, and the nodes corresponding to the maximum
         EC value of each column are selected in turn.

20:   A.add(N)

21:    **End if**

22:   **End for**

23:   $t = $ getTime();

24:   **If** $(t \geq T)$

25:   $M^T \leftarrow getUpdateNewMeetMatrix(M^T, R^T, S^T, T)$ //After the Δt time information is passed, the
         updated target encounter probability matrix is used as the input for calculating the encounter
         probability matrix in the next round of information transfer.

Return Step 5;

26:   **End if**

Algorithm 2: Matrix update algorithm

Input: M The matrix trained by Algorithm 1

$M^{t_1} \cdots M^{t_n}$ The encounter matrixes carried by the nodes entering the area that are received at different time intervals.

Output: M^T Complete the updated matrix

1:     **Let** $k \in [1, n]; i, j \in [1, z]$//$n$ is the total number of encounter matrices transmitted by nodes in the area at different time periods, z is the maximum number of rows or columns of the matrix M^{t_k}

2:     **For** $k = n$ to 1 do//The later the matrix arrives, the more it will be processed first.

3:     **If** matrix M does not contain the node i in the matrix M^{t_k} **do**//If the matrix M does not contain a row or a column.

4:     $M.add(M^{t_k}.getRow(i), M^{t_k}.getColumn(i))$//Add this row or column directly

5:     **Else if** M_{ij} is null and $M_{ij}^{t_k}$ is not null//If the value of the i-th row and the j-column in the matrix M is null, and the value of the i-th row and the j-column of the matrix M^{t_k} reached at time t_k is not null, the value is filled into the matrix M.

6:     $M_{ij}.fill(M_{ij}^{t_k})$

7:     **Else if** M_{ij} is not null and $M_{ij}^{t_k}$ is not null//If the value of the i-th row and the j-th column in the matrix is not null, and the value of the i-th row and the j-th column of the matrix reached at time t_k is not null

8:     **If** $i \notin M.getHisArea()$ and $j \in M.getHisArea()$//If node i does not belong to the area and j belongs to the area, the value of M_{ij} is updated to $M_{ij}^{t_k}$.

9:     $M_{ij}.update(M_{ij}^{t_k})$

10:     **End if**

11:     **End if**

12:     **End for**

13:     **If** $(i = j)$//The matrix diagonal value is set to 1.

14:     $M_{ij} = 1$

15:     **End if**

16:     **Output** M^T//M^T is the updated encounter probability matrix.

3.6. Complexity Analysis

The main calculation method of the gradient method is to calculate the objective function, L, and its gradient according to the variables. The computational complexity of the objective function, L, is $O(\rho_M l + \rho_M \bar{v} l + \rho_R l)$, where ρ_M is the number of non-zero terms in the matrix, M, and $\bar{v}$ is the average number of nodes that have a social relationship with the node. Since there are not many users in a region who have a social relationship value with a user, this indicates that the value of $\bar{v}$ is relatively small. The computational complexities for the gradients, $\frac{\partial L}{\partial N_i}$, $\frac{\partial L}{\partial X_j}$, and $\frac{\partial L}{\partial Y_k}$, in Equation (25) are $O(\rho_M \bar{s} l + \rho_M \bar{s} \bar{v} l + \rho_R l)$, $O(\rho_M l + \rho_M \bar{v} l)$ and $O(\rho_R l)$, respectively. Among them, $\bar{s}$ is the average number of devices with social relationships that are close to one device, ρ_R is the number of non-zero terms in the matrix, R. Actually, in a complex social network, the value of $\bar{s}$ is approximately equal to the value of $\bar{v}$. Therefore, the total computational complexity of an iteration is $O(\rho_M \bar{s} l + \rho_M \bar{s} \bar{v} l + \rho_R l)$, which indicates the calculation time of our method is linear with the amount of information collected by the node. If the obtained matrices are all sparse matrices, the algorithm will execute rapidly with a time complexity of $O(n)$. It also applies when the number of nodes is large. However, if the obtained matrix is a dense matrix, the complexity in the communication range of big data with dense nodes and complex relations is difficult to estimate. Since the complexity of the CAOP, SSIS, and SIaOR algorithms are both $O(n^2)$, the complexity analysis shows that the proposed method is simpler than them.

4. Experiment Analysis

In this section, we evaluate the performance of the proposed model, PEBN, in OMNets. The OMNets is a free, open source multi-protocol network simulation software, which plays an important role in network simulation.

4.1. Parameter Settings

The article uses the open source simulation tool, OMNets, to make the simulation. The network has a size of 3400 m * 4500 m and 100–1300 nodes. Each node is moving randomly. Their moving speeds are, respectively, 1 m/s and 10 m/s. After reaching its destination, the node stays there for a random time. The sensing radius of the node is 50 m. The message generation interval is 25 to 35 s, the transmission interval is 17–18 min. The Time to Live is set to 100 min. The matrix update interval is 180 min. In the above simulation environment, we compared the performance of CAOP, SSIS, and SIaOR algorithms and our proposed routing algorithm (PEBN) in the delivery rate, network overhead rate, and hop count with the change of the number of nodes and simulation time.

4.2. Parameter Analysis

The performance of PEBN is affected by some time-dependent parameters. First, we analyzed the parameters before comparing the algorithms. In order to improve the performance of the algorithm, we set the reasonable parameter values by experimentally analyzing the update period of the matrix and the node message delivery time interval.

Different time periods, T, have different effects on the results of model training. Too short a time period will increase the network resource consumption. If the time period is too long, it will not be able to timely capture the latest dynamic changes of the node, reducing the accuracy of prediction, and affecting the transmission success rate. Therefore, we need to select the appropriate time period to update the predicted target matrix. Through analysis, as shown in Figure 3a, we can observe the fluctuation of the curve. As the time period increases, the transmission success rate increases first and then decreases. The transmission success rate reaches a large value at 175 min. Due to the change of the time period, the average transmission hop count is also affected. If the time period is set too long, it will affect the value of the model training and will be not accurate enough, resulting in more hops. Therefore, it can be seen from Figure 3b that as the time period increases, the hop count decreases and then increases. The phenomenon is because the time period is too short to collect enough information to reduce the error. When the time period reaches 200 min, the hop count is small. We can use the tradeoff method to determine that the period, T, is 180 min and increase the number of hops to ensure the delivery ratio.

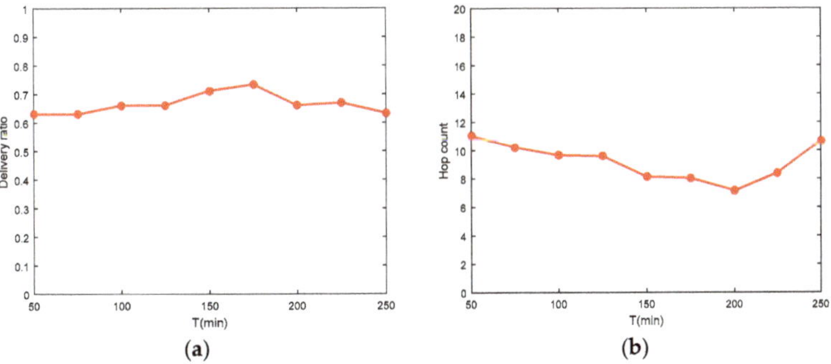

Figure 3. The relationship between delivery ratio and average hop count with time periods T: (**a**) delivery ratio with different T; (**b**) average hop count with different T.

We set the node to pass a message at Δt intervals. When a node transmits a message, a copy of the message is sent from a node within the messaging range that filters EC values greater than those of itself and the destination node. If the interval, Δt, is too short, it will cause the nodes in the network to send the same message copies to the same neighbor node more than once. The neighbor node receives the same message copy and refuses to receive the message, resulting in unnecessary network overhead. If the time interval, Δt, is too long, it will cause the message in the network to spread slowly and increase the transmission delay of the message. From Figure 4b, we can observe that as Δt increases, the number of nodes covered by message copies transmitted by the node also decreases. During the 16 min to 18 min time period, the number of nodes covered by the message copy is large. However, the transmission of messages after Δt interval also affects the transmission success rate of the nodes. Just because a larger number of nodes are covered by the message does not mean that the delivery ratio can be improved; it may cause route congestion and cause information loss. Therefore, we can observe from Figure 4a that when Δt reaches 20 min, our delivery rate reaches the highest value. Since the value of Δt is between 16 min and 18 min, the node coverage does not fluctuate, and the transmission success rate fluctuates less in the range of 18–20 min. Therefore, we set Δt to 18 min.

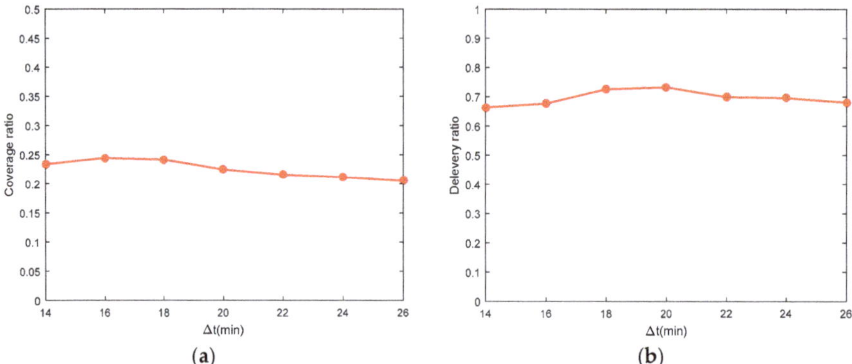

Figure 4. The relationship between delivery ratio and node coverage ratio with different time interval Δt: (**a**) node coverage ratio different Δt; (**b**) delivery ratio with different Δt.

To sum up, in the model with consideration of the node's transmission success rate and transmission delay variation, we set $T = 180$ min and $\Delta t = 18$ min for the following experiments.

4.3. Analysis of Simulation Results

4.3.1. Methodology

In a real environment, each node is selfish. To make the simulation experiment closer to the real scene, we randomly set the selfishness of the node relative to other nodes in the experiment. When a node forwards a message, it forwards the message according to a certain cooperation probability that removes the selfishness. It should be emphasized that we aim to select nodes that have a high probability of encountering the target node and are willing to cooperate to forward the message. In the experiment, the cooperation matrix of the algorithm was not set manually, but was based on the cooperative probability calculated by the number of times the node helps a node to forward the message in the previous T period.

The evaluation of the model mainly comes from two aspects: (1) Effectiveness analysis. The experiment compares the difference between our presented model and other existing models in optimizing the network structure and improving the information delivery ratio. (2) Adaptability analysis: The probability matrix is periodically updated by various dynamic change information, and the nodes that can cooperate and have a high probability of encountering with the destination node are filtered to transmit data, thereby increasing the delivery ratio and reducing the average hop count. As a reference, we set a list of performance metrics, which is used for comparison, as follows:

1. Delivery ratio is the ratio of the number of delivered messages to the total number of generated messages.
2. Overhead is the average number of intermediate nodes used for one delivered message.

3. Hop count is the average hops of successfully delivered messages.

We use these performance metrics to analyze the effects of the model and compare them with other researchers' proposed solutions. To compare the performance of the PEBN protocol with other schemes, the delivery ratio, overhead ratio, and hop count were measured.

4.3.2. Impacts on Delivery Ratio

Figures 5 and 6 shows the delivery performance of the PEBN protocol when the number of nodes and simulation time are different, respectively. From Figure 5 we can see that the delivery ratio of messages by the PEBN protocol is higher than other schemes due to better choices while sending the messages. This is because our solution sets a different degree of selfishness to each node for other nodes. Other schemes can easily lead to message loss, because they cannot fully consider the cooperating factors and social relationship factors of the nodes, and even cannot get more information about the nodes. Therefore, our algorithm can obtain more accurate predictions to have a better choice in the process of screening nodes. Moreover, as the number of nodes increases, the training of the prediction matrix is more extensive and accurate due to the increase of node social information. Thus, obtaining more unknown information between nodes, increasing the number of candidate nodes in the information transmission process, and improving the accuracy of node selection, thus makes it more likely to select the optimal node. From Figure 6, it can be observed that the delivery performance of the PEBN protocol is better than the other scheme. Because we not only consider the probability of encountering the target node, but also the path of the target node, it is unlikely to lose the packet. In addition, as the simulation time increases, the message delivery rate also increases, and the message will have more time to wait for the next best node.

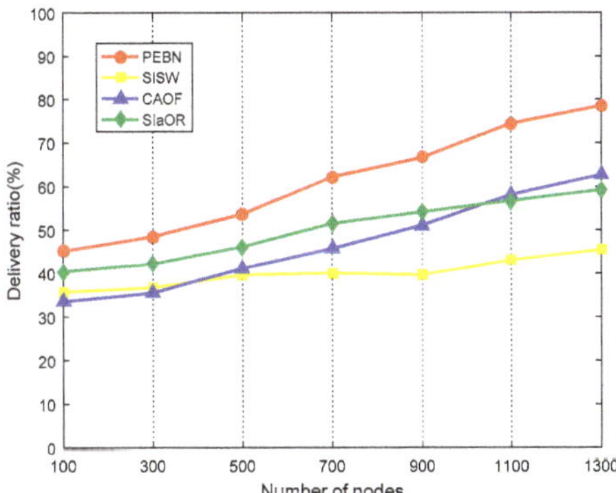

Figure 5. The relationship between the number of nodes and the delivery ratio.

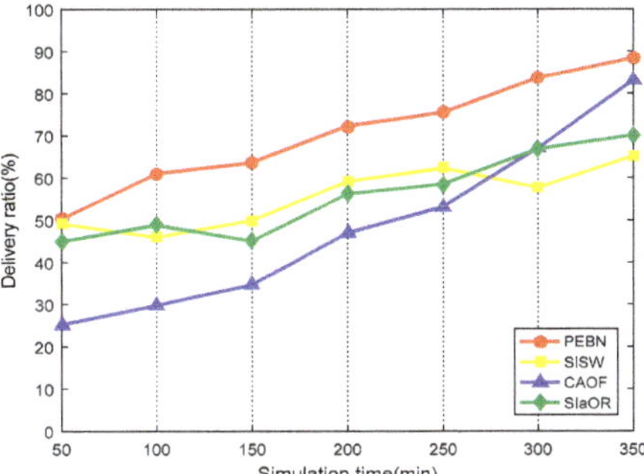

Figure 6. The relationship between time and delivery ratio.

4.3.3. Impacts on Average Hop Count

Figures 7 and 8 shows the hop count of the PEBN protocol when the number of node and simulation time varies, respectively. In Figure 7, the average hop count of the protocols is compared with others schemes. The PEBN protocol has the smallest average hop count in the compared scheme because it optimizes the routing of message transmissions more efficiently than other schemes. It can be seen in Figure 7 that as the number of nodes increases, the average hop count of the node increases and then gradually stabilizes. Our solution considers many factors, and as the number of nodes increases, the relationship between nodes becomes more complex, and more hidden features between nodes are mined through probabilistic prediction models, thus reducing the number of hops from the source node to the destination node. However, other methods may not have many parameters to enhance the adaptability of the model. Therefore, our model performs better. From Figure 8, it can be observed that our proposed scheme has fewer hops than other schemes. This is because our solution collects and processes node feature information, so we can get as much data as possible to more accurately predict the probability of encounters between nodes. As the simulation time increases, the node can capture more information about the movement of the nodes, which can add more additional information to the process of selecting the key nodes, making the model more robust. Therefore, the average hop count of the node is gradually reduced and stabilized.

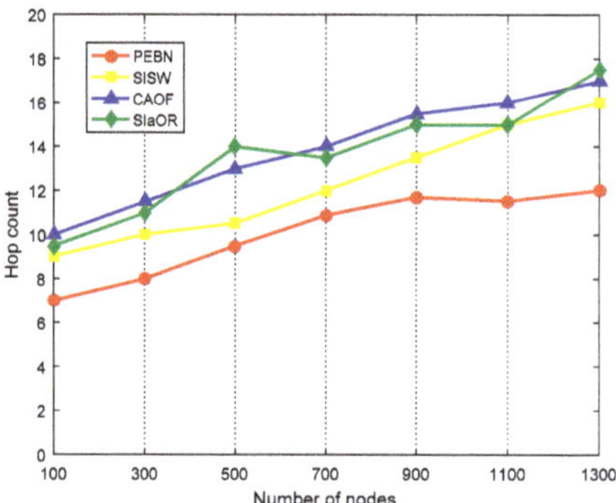

Figure 7. The relationship between the number of nodes and hop count.

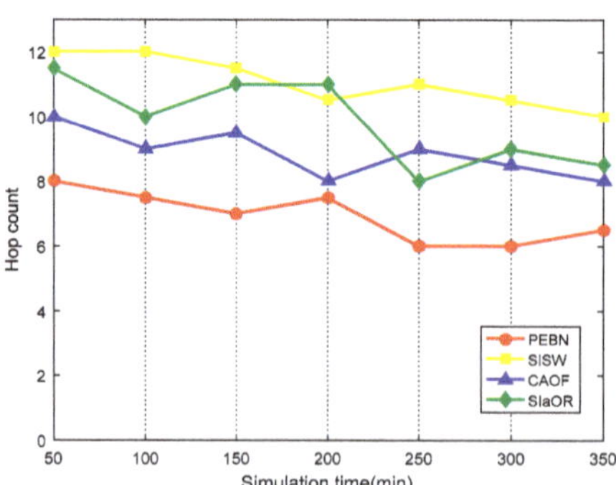

Figure 8. The relationship between time and hop count.

4.3.4. Impacts on Routing Overhead

Figures 9 and 10 shows the routing overhead of the PEBN protocol when the number of nodes and simulation time are varied, respectively. From Figure 9, it can be observed that the routing overhead of our scheme is compared with other schemes. Compared with the result, PEBN has a better performance than other models. Since we apply the node cooperation to our proposed model, we can also perform well in scenarios where the node has selfish features. PNEC can predict transmitting neighbors better than other models; the message will be forwarded to a node with a high probability of satisfying the destination in the transmission route, which effectively reduces the cost loss of sending the message to the non-cooperative node. In addition, we can also observe that as the number of nodes increases, the amount of routing overhead also increases. An increase in the number of nodes means the number of candidate nodes for message selection increases. Messages need to be passed to more nodes, thereby increasing the overhead of information being propagated over the network. Moreover, this same observation can be seen in Figure 10, which shows that routing overhead increases as the simulation time is increased. Because the nodes do not need to continuously calculate and decide through the proposed model during the information transmission process, the transmission task of nodes becomes simpler. Furthermore, we have more complex probability models to choose neighbors, so our overhead increase ratio is relatively smaller than other models with the increase of simulation time.

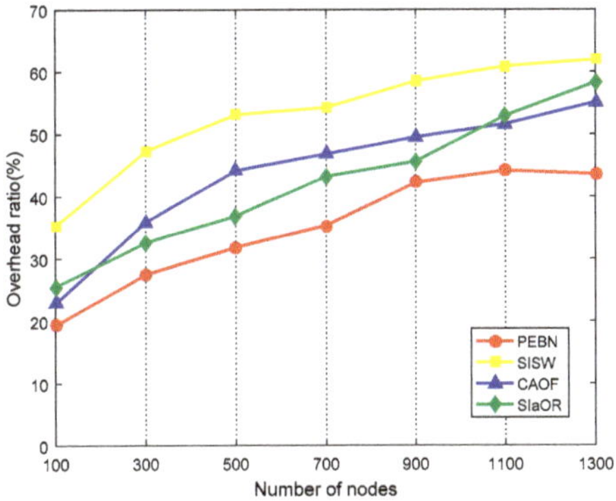

Figure 9. The relationship between the number of nodes and overhead.

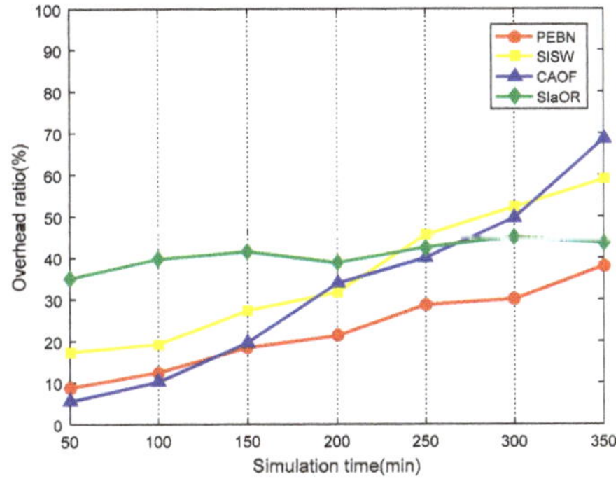

Figure 10. The relationship between time and overhead.

The conclusion can be drawn by comparing the three indicators of several schemes with the number of nodes and the simulation time. The PEBN algorithm adds social relationships and cooperation relationships when predicting probabilities so that a node can deliver a message to a cooperating node that has a high probability of encountering a destination node. Experiments show that the PEBN algorithm outperforms other algorithms in terms of the transmission success rate, routing overhead, and hop count. Specifically, the algorithm not only improves the data transmission efficiency in the network, but also adopts the updated information to adapt to the current network environment when the network topology changes dynamically. In addition, handling complex transmission decisions through base stations or edge devices reduces the node overhead. By combining the social and cooperative relationships between nodes with a probabilistic prediction model, a better transmission path can be selected to reduce the number of hops to the destination node. Therefore, our scheme can improve performance in the transmission environment.

5. Conclusions

We proposed a routing decision method based on an improved probability model combined with a quantitative social relationship value and cooperative value to filter neighbor nodes. The algorithm combines multiple feature information between nodes and uses this feature information to quantify social relationship values and partnership values. Then, the prediction matrix was obtained by matrix decomposition and gradient descent, and the relay nodes were filtered according to the predicted probability values. In our model, we first quantified the node social relationship value and the cooperative relationship value based on the collected information to form the social relationship matrix and the cooperation relationship matrix. Then, we used them to update and predict the probability of encountering cooperation between nodes in the way we proposed. Finally, in the transmission phase, the node requested a probability table associated with the destination node, and selected a node with a high probability of encountering the destination node as the next hop node. The simulation results show that the protocol performs better than the SISW, CAOF, and SIaOR transmission models in the transmission success rate, average hop count, and overhead. For a single node, the model optimizes the path from the source node to the target node. For the entire network, the performance of the network is improved to accommodate large-scale data transmission in the 5G. In the future, we will use real data sets to simulate real-world scenarios and explore other more efficient ways to improve information transmission.

Author Contributions: G.Y., Z.C. and J.W. (Jia Wu) conceived the idea of the paper. G.Y., Z.C., J.W. (Jia Wu) and J.W. (Jian Wu) drafted the manuscript and collected the data, wrote the code and performed the analysis; Z.C. contributed reagents/materials/analysis tools; G.Y. wrote and revised the paper.

Funding: This research was funded by [The Major Program of National Natural Science Foundation of China] grant number [No. 71633006]; [The National Natural Science Foundation of China] grant number [No. 616725407] [No. 61379057]; [China Postdoctoral Science Foundation funded project] grant number [2017M612586]; [The Postdoctoral Science Foundation of Central South University] grant number [185684].

Acknowledgments: This work was supported partially by "Mobile Health" Ministry of Education—China Mobile Joint Laboratory.

Conflicts of Interest: The authors declare that they have no competing interests.

References

1. Wu, J.; Chen, Z.; Zhao, M. Effective information transmission based on socialization nodes in opportunistic networks. *Comput. Netw.* **2017**, *129*, 297–305. [CrossRef]
2. Wu, J.; Chen, Z.; Zhao, M. Information cache management and data transmission algorithm in opportunistic social networks. *Wirel. Netw.* **2018**, 1–12. [CrossRef]
3. Wu, J.; Chen, Z.; Zhao, M. Weight Distribution and Community Reconstitution Based on Communities Communications in Social Opportunistic Networks. *Peer Peer Netw. Appl.* **2018**, 1–9. [CrossRef]
4. Zhao, Y.; Song, W.; Han, Z. Social-Aware Data Dissemination via Device-to-Device Communications: Fusing Social and Mobile Networks with Incentive Constraints. *IEEE Trans. Serv. Comput.* **1939**. [CrossRef]
5. Lokhov, A.Y.; Mézard, M.; Ohta, H.; Zdeborová, L. Inferring the Origin of an Epidemic with Dynamic Message-Passing Algorithm. *Phys. Rev. E Stat. Nonlinear Soft Matter Phys.* **2014**, *90*, 012801. [CrossRef] [PubMed]
6. Korkmaz, T.; Krunz, M. *A Randomized Algorithm for Finding a Path Subject to Multiple QoS Requirements*; Elsevier North-Holland, Inc.: Amsterdam, The Netherlands, 2001.
7. Aliotta, J.M.; Pereira, M.; Johnson, K.W.; de Paz, N.; Dooner, M.S.; Puente, N.; Ayala, C.; Brilliant, K.; Berz, D.; Lee, D.; et al. Microvesicle Entry into Marrow Cells Mediates Tissue-Specific Changes in Mrna by Direct Delivery of Mrna and Induction of Transcription. *Exp. Hematol.* **2010**, *38*, 233–245. [CrossRef] [PubMed]

8. Wu, J.; Zhao, M.; Chen, Z. Small Data: Effective Data Based on Big Communication Research in Social Networks. *Wirel. Pers. Commun.* **2018**, *99*, 1–14. [CrossRef]

9. Chuang, Y.J.; Lin, C.J. Cellular Traffic Offloading through Community-Based Opportunistic Dissemination. In Proceedings of the Wireless Communications and NETWORKING Conference, Shanghai, China, 1–4 April 2012.

10. Cheng, H.H.; Lin, C.J. Source Selection and Content Dissemination for Preference-Aware Traffic Offloading. *IEEE Trans. Parallel Distrib. Syst.* **2015**, *26*, 3160–3174. [CrossRef]

11. Ning, T.; Yang, Z.; Wu, H.; Han, Z. Self-Interest-Driven Incentives for Ad Dissemination in Autonomous Mobile Social Networks. In Proceedings of the IEEE INFOCOM, Turin, Italy, 14–19 April 2013.

12. Liu, G.; Ji, S.; Cai, Z. Strengthen Nodal Cooperation for Data Dissemination in Mobile Social Networks. *Pers. Ubiquitous Comput.* **2014**, *18*, 1797–1811. [CrossRef]

13. Tsiropoulou, E.E.; Mitsis, G.; Papavassiliou, S.; Tsiropoulou, E.E.; Mitsis, G.; Papavassiliou, S. Interest-Aware Energy Collection & Resource Management in Machine to Machine Communications. *Ad Hoc Netw.* **2018**, *68*, 48–57.

14. Tsiropoulou, E.E.; Paruchuri, S.T.; Baras, J.S. Interest, Energy and Physical-Aware Coalition Formation and Resource Allocation in Smart Iot Applications. In Proceedings of the Information Sciences and Systems, Baltimore, MD, USA, 22–24 March 2017.

15. Tsiropoulou, E.E.; Koukas, K.; Papavassiliou, S. A Socio-Physical and Mobility-Aware Coalition Formation Mechanism in Public Safety Networks. *EAI Endorsed Trans. Future Internet* **2018**, *4*, 154176. [CrossRef]

16. Lin, C.R.; Gerla, M. Adaptive Clustering for Mobile Wireless Networks. *IEEE J. Sel. Areas Commun.* **1997**, *15*, 1265–1275. [CrossRef]

17. Liu, W.; Gerla, M. Routing in Clustered Multihop Mobile Wireless Networks with Fading Channel. In Proceedings of the IEEE SICON, Singapore, 14–17 April 1997.

18. Gerla, M.; Tsai, T.C. Multicluster, Mobile, Multimedia Radio Network. *Wirel. Netw.* **1995**, *1*, 255–265. [CrossRef]

19. Wang, G.; Zheng, L.; Yan, L.; Zhang, H. Probabilistic routing algorithm based on transmission capability of nodes in DTN. In Proceedings of the 2017 11th IEEE International Conference on Anti-counterfeiting, Security, and Identification (ASID), Xiamen, China, 27–29 October 2017; pp. 146–149.

20. Zeydan, E.; Bastug, E.; Bennis, M.; Kader, M.A.; Karatepe, I.A.; Er, A.S.; Debbah, M. Big Data Caching for Networking: Moving from Cloud to Edge. *IEEE Commun. Mag.* **2016**, *54*, 36–42. [CrossRef]

21. Tao, J.; Wu, H.; Shi, S.; Hu, J.; Gao, Y. Contacts-aware opportunistic forwarding in mobile social networks: A community perspective. In Proceedings of the 2018 IEEE Wireless Communications and Networking Conference (WCNC), Barcelona, Spain, 15–18 April 2018; pp. 1–6.

22. Wang, R.; Wang, X.; Hao, F.; Zhang, L.; Liu, S.; Wang, L.; Lin, Y. Social Identity–Aware Opportunistic Routing in Mobile Social Networks. *Trans. Emerg. Telecommun. Technol.* **2018**, *29*, e3297. [CrossRef]

23. Jia, B.; Zhou, T.; Li, W.; Xu, Z. An Opportunity Transmission Mechanism in Mobile Crowd Sensing Network based on SSIS Model. In Proceedings of the 2018 IEEE 22nd International Conference on Computer Supported Cooperative Work in Design (CSCWD), Nanjing, China, 9–11 May 2018; pp. 695–700.

24. Sharma, D.K.; Dhurandher, S.K.; Woungang, I.; Srivastava, R.K.; Mohananey, A.; Rodrigues, J.J.P.C. A Machine Learning-Based Protocol for Efficient Routing in Opportunistic Networks. *IEEE Syst. J.* **2016**, *12*, 2207–2213. [CrossRef]

25. Salakhutdinov, R.; Mnih, A. Probabilistic Matrix Factorization. In Proceedings of the International Conference on Neural Information Processing Systems, Vancouver, BC, Canada, 3–6 December 2007.

26. Dueck, D.; Frey, B.J. *Probabilistic Sparse Matrix Factorization*; University of Toronto Technical Report Psi; University of Toronto: Toronto, ON, Canada, 2004.

Article

Derivative Free Fourth Order Solvers of Equations with Applications in Applied Disciplines

Ramandeep Behl [1], **Ioannis K. Argyros** [2], **Fouad Othman Mallawi** [1] **and J. A. Tenreiro Machado** [3,*]

[1]   Department of Mathematics, King Abdulaziz University, Jeddah 21589, Saudi Arabia;
      ramanbehl87@yahoo.in (R.B.); rlal@kau.edu.sa (F.O.M.)
[2]   Department of Mathematics Sciences, Cameron University, Lawton, OK 73505, USA; ioannisa@cameron.edu
[3]   ISEP-Institute of Engineering, Polytechnic of Porto Department of Electrical Engineering,
      431 4294-015 Porto, Portugal
*   Correspondence: jtm@isep.ipp.pt

Received: 27 March 2019; Accepted: 17 April 2019; Published: 23 April 2019

Abstract: This paper develops efficient equation solvers for real- and complex-valued functions. An earlier study by Lee and Kim, used the Taylor-type expansions and hypotheses on higher than first order derivatives, but no derivatives appeared in the suggested method. However, we have many cases where the calculations of the fourth derivative are expensive, or the result is unbounded, or even does not exist. We only use the first order derivative of function Ω in the proposed convergence analysis. Hence, we expand the utilization of the earlier scheme, and we study the computable radii of convergence and error bounds based on the Lipschitz constants. Furthermore, the range of starting points is also explored to know how close the initial guess should be considered for assuring convergence. Several numerical examples where earlier studies cannot be applied illustrate the new technique.

Keywords: divided difference; radius of convergence; Kung–Traub method; local convergence; Lipschitz constant; Banach space

MSC: 47J05; 47J25; 65H10; 65G99

---

1. Introduction

We look for a unique root p_* of the equation:

$$\Omega(v) = 0, \tag{1}$$

where Ω is a continuous operator defined on a convex subset $\mathbb{P}$ of $\mathbb{S}$ with values in $\mathbb{S}$, and $\mathbb{S} = \mathbb{R}$ or $\mathbb{S} = \mathbb{C}$. This is a relevant issue since several problems from mathematics, physics, chemistry, and engineering can be reduced to Equation (1).

In general, either the lack, or the intractability of analytic solutions force researchers to adopt iterative techniques. However, when using that type of approach, we find problems such as slow convergence, converge to undesired root, divergence, computational inefficiency, or failure (see Traub [1] and Petković et al. [2]). The study of the convergence of iterative algorithms can be classified into two categories, namely the semi-local and local convergence analysis. The first case is based on the information in the neighborhood of the starting point. This also gives criteria for guaranteeing the convergence of iteration algorithms. Therefore, a relevant issue is the convergence domain, as well as the radii of convergence of the algorithm.

Herein, we deal with the second case, that is the local convergence analysis. Let us consider a fourth order algorithm defined for $n = 0, 1, 2, \ldots$, as:

$$\lambda_s = \delta_s + \beta\Omega(\delta_s)^k, \text{ with } \beta \neq 0 \in \mathbb{R},$$

$$\mu_s = \lambda_s - \frac{\Omega(\lambda_s)}{[\delta_s, \lambda_s; \Omega]},$$

$$\delta_{s+1} = \mu_s - H(v_s, w_s)\frac{\Omega(\mu_s)}{[\delta_s, \lambda_s; \Omega]},$$

$$(2)$$

where $\lambda_0 \in \mathbb{P}$ is an initial point, $k \in \mathbb{N}$ (k is an arbitrary natural number), $[\delta_s, \lambda_s; \Omega] : \mathbb{P} \times \mathbb{P} \rightarrow L(\mathbb{S}, \mathbb{S})$ satisfies $[\delta_s, \lambda_s; \Omega] = \frac{\Omega(x) - \Omega(y)}{x-y}$ for $x \neq y$, $v_s = \frac{\Omega(\mu_s)}{\Omega(\lambda_s)}$, $w_s = \frac{\Omega(\mu_s)}{\Omega(\delta_s)}$, and $H : \mathbb{S} \times \mathbb{S} \rightarrow \mathbb{S}$ is a continuous function. The fourth order convergence for Method (2) was studied by Lee and Kim [3] with Taylor series, hypotheses up to the fourth order derivative of function Ω, and hypotheses on the first and second partial derivatives of function H. However, only the divided difference of the first order appears in (2). Favorable computations were also given with related Kung–Traub methods [1] of the form:

$$\lambda_s = \delta_s + \beta\Omega(\delta_s)^4, \text{ with } \beta \neq 0 \in \mathbb{R},$$

$$\mu_s = \lambda_s - \frac{\Omega(\lambda_s)}{[\delta_s, \lambda_s; \Omega]},$$

$$\delta_{s+1} = \mu_s - \frac{\Omega(\delta_s)}{\Omega(\delta_s) - 2\Omega(\mu_s)}\frac{\Omega(\mu_s)}{[\lambda_s, \mu_s; \Omega]}.$$

$$(3)$$

Notice that (3) is obtained from (2), if we define function H as $H(v, w) = \frac{1}{1-2w}$. The assumptions on the derivatives of Ω and H restrict the suitability of Algorithms (2) and (3). For instance, let us consider Ω on $\mathbb{P} = \mathbb{S} = \mathbb{R}$, $\mathbb{P}_1 = [-\frac{1}{\pi}, \frac{2}{\pi}]$ as:

$$\Omega(v) = \begin{cases} v^3 \log(\pi^2 v^2) + v^5 \sin\left(\frac{1}{v}\right), & v \neq 0 \\ 0, & v = 0 \end{cases}.$$

From this expression, we obtain:

$$\Omega'(v) = 2v^2 - v^3 \cos\left(\frac{1}{v}\right) + 3v^2 \log(\pi^2 v^2) + 5v^4 \sin\left(\frac{1}{v}\right),$$

$$\Omega''(v) = -8v^2 \cos\left(\frac{1}{v}\right) + 2v(5 + 3\log(\pi^2 v^2)) + v(20v^2 - 1)\sin\left(\frac{1}{v}\right),$$

$$\Omega'''(v) = \frac{1}{v}\left[(1 - 36v^2)\cos\left(\frac{1}{v}\right) + v\left(22 + 6\log(\pi^2 v^2) + (60v^2 - 9)\sin\left(\frac{1}{v}\right)\right)\right].$$

We find that $\Omega'''(v)$ is unbounded on $\mathbb{P}_1$ at the point $v = 0$. Therefore, the results in [3] cannot be applied for the analysis of the convergence of Methods (2) or (3). Notice that there are numerous algorithms and convergence results available in the literature [1–15]. Nonetheless, practice shows that the initial prediction must be in the neighborhood of the root for achieving convergence. However, how close must it be to the starting point? Indeed, local results do not give any information about the ball convergence radii.

We broaden the suitability of Methods (2) and (3) by using only assumptions on the first derivative of function Ω. Moreover, we estimate the computable radii of convergence and the error bounds from Lipschitz constants. Additionally, we discuss the range of initial estimate p_* that tells us how close it must be to achieve a granted convergence of (2). This problem was not addressed in [3], but is of capital importance in practical applications.

In what follows: Section 2 addresses the study of local convergence (2) and (3). Section 3 contains three numerical examples that illustrate the theoretical formulation. Finally, Section 4 gives the concluding remarks.

2. Convergence Analysis

Let $b > 0$, $\alpha > 0$, $\gamma > 0$, $\beta \in \mathbb{S}$, $k \in \mathbb{N}$ and $M \geq 1$ be given constants. Furthermore, we consider that $H : \mathbb{S} \times \mathbb{S} \to \mathbb{S}$, $h : [0, \infty) \to [0, \infty)$ are continuous functions such that:

$$|H(v, \eta)| \leq |H(|v|, |\eta|)| \leq h(v), \tag{4}$$

for each $v, \eta \in \mathbb{S}$ with $|\eta| \leq v$, and that $|H|$ and h are nondecreasing functions on the interval $\left[0, \frac{1}{\gamma}\right)^2$, $\left[0, \frac{1}{\gamma}\right)^2$, respectively. For the local convergence analysis of (2), we need to introduce a few functions and parameters. Let us define the parameters R_0 and R_1 given by:

$$R_0 = \frac{1}{(1+\alpha)\gamma}, \quad R_1 = \frac{1}{(1+\alpha)\gamma + \gamma\alpha(b|\beta|M + \alpha)}, \tag{5}$$

and function g_1 on the interval $[0, R_1)$ by:

$$g_1(v) = \frac{\gamma\alpha(b|\beta|M + \alpha)v}{1 - (1+\alpha)\gamma v}. \tag{6}$$

From the above functions, it is easy to see that $R_1 < R_0 < \frac{1}{\gamma}$, $g_1(R_1) = 1$ and $0 \leq g_1(v) < 1$, for $v \in [0, R_1)$. Moreover, we consider the functions q and $\bar{q}$ on $[0, R_1)$ as:

$$q(v) = \gamma(\alpha + g_1(v))v \quad \text{and} \quad \bar{q}(v) = q(v) - 1.$$

It is straightforward to find that $\bar{q}(0) = -1 < 0$ and that $\bar{q}(v) \to +\infty$ as $v \to r_1^-$. By the intermediate value theorem, we know that $\bar{q}$ has zeros in the interval $(0, R_1)$. Let us assume that R_q is the smallest zero of function $\bar{q}$ on $(0, R_1)$, and set:

$$\bar{r} = \min\{R_1, R_q\}. \tag{7}$$

Furthermore, let us define functions g_2 and $\bar{g}_2$ on $[0, \bar{r})$ such that:

$$g_2(v) = \left(1 + \frac{Mh(v)}{1 - q(v)}\right) g_1(v) \tag{8}$$

and:

$$\bar{g}_2(v) = g_2(v) - 1. \tag{9}$$

Suppose that:

$$\bar{g}_2(v) \to \text{ a positive number or } +\infty, \text{ as } v \to \bar{r}^-. \tag{10}$$

From (8), we have that $\bar{g}_2(0) < 0$ and from (10) that $\bar{g}_2(v) > 0$ as $v \to \bar{r}^{-1}$. Further, we assume that R is the smallest zero of function $\bar{g}_2$ on $(0, \bar{r})$. Therefore, we have that for each $v \in [0, r)$:

$$0 \leq g_1(v) < 1, \tag{11}$$

$$0 \leq g_2(v) < 1, \tag{12}$$

$$0 \leq q(v) < 1. \tag{13}$$

Let us denote by $U(\mu,\ r)$ and $\bar{U}(\mu,\ r)$ the open and closed balls in $\mathbb{S}$ with center $\mu \in S$ and of radius $r > 0$, respectively.

Theorem 1. *Let us assume that* $\Omega : \mathbb{P} \subset \mathbb{S} \to \mathbb{S}$ *is a differentiable function and* $[\cdot,\cdot\ ;\Omega] : \mathbb{P} \times \mathbb{P} \to L(\mathbb{S},\ \mathbb{S})$ *is a divided difference of first order of* Ω. *Furthermore, we consider that* h *and* H *are functions satisfying* (4), (9), $p_* \in \mathbb{P}$, $b > 0$, $\alpha > 0$, $\gamma > 0$, $M \geq 1, k \in \mathbb{N}$, $\beta \in S$ *and that for each* $x,\ y \in \mathbb{P}$, *we have:*

$$\Omega(p_*) = 0, \quad \Omega'(p_*) \neq 0, |\Omega'(p_*)| \leq b, \tag{14}$$

$$|\Omega'(p_*)^{-1}([x,\ y,\ \Omega] - \Omega'(p_*)| \leq \gamma(|x - p_*| + |y - p_*|), \tag{15}$$

$$h(v) = H\left(\frac{M\gamma(|\beta|Mb + \alpha)v}{(1 - \gamma\alpha v)(1 - \gamma(1 + \alpha)v)},\ \frac{Mg_1(v)}{1 - \gamma v}\right) \tag{16}$$

$$|I + \beta[x,\ p_*;\Omega]^k(x - p_*)^{k-1}| \leq \alpha, \tag{17}$$

$$|\Omega'(p_*)^{-1}[x,\ p_*,\ \Omega]| \leq M, \tag{18}$$

$$\bar{U}(p_*, \alpha r) \subseteq \mathbb{P}. \tag{19}$$

Then, the sequence $\{\delta_s\}$ *obtained for* $\lambda_0 \in U(p_*,\ R) - \{x^*\}$ *by* (2) *is well defined, remains in* $U(p_*,\ R)$ *for each* $n = 0, 1, 2, \ldots$, *and converges to* p_*, *so that:*

$$|\lambda_s - p_*| \leq \alpha|\delta_s - p_*| < R, \tag{20}$$

$$|\mu_s - p_*| \leq g_1(|\delta_s - p_*|)|\delta_s - p_*| \leq |\delta_s - p_*| < R, \tag{21}$$

$$|\delta_{s+1} - p_*| \leq g_2(|\delta_s - p_*|)|\delta_s - p_*| < |\delta_s - p_*|, \tag{22}$$

and $G \in [R, \frac{1}{\gamma})$. *Moreover, the limit point* p_* *is the unique root of equation* $\Omega(x) = 0$ *in* $\bar{U}(p_*,\ G) \cap \mathbb{P}$.

Proof. By hypotheses $\lambda_0 \in U(p_*, r) - \{x^*\}$, (14), (17) and (19), we further obtain:

$$\delta_0 - p_* = \lambda_0 - p_* + \beta\left(\Omega(\lambda_0) - \Omega(p_*)\right)^k$$
$$= \left(I + \beta[\lambda_0,\ p_*;\Omega]^k(\lambda_0 - p_*)^{k-1}\right)(\lambda_0 - p_*),$$

so that:

$$|\delta_0 - p_*| = \left|I + \beta[\lambda_0,\ p_*;\ \Omega]^k(\lambda_0 - p_*)^{k-1}\right||\lambda_0 - p_*|$$
$$\leq \alpha r, \tag{23}$$

which leads to (20) for $s = 0$ and $\delta_0 \in U(p_*,\ \alpha r)$. We need to show that $[\lambda_0,\ \delta_0;\ \Omega] \neq 0$. Using (15) and the definition of R, we obtain:

$$\left|\Omega'(p_*)^{-1}([\lambda_0,\ \delta_0;\ \Omega] - \Omega'(p_*)\right| \leq \gamma\left(|\lambda_0 - p_*| + |\delta_0 - p_*|\right)$$
$$\leq \gamma\left(|\lambda_0 - p_*| + \alpha|\lambda_0 - p_*|\right) \tag{24}$$
$$\leq \gamma(1 + \alpha)|\lambda_0 - p_*| < \gamma(1 + \alpha)R < 1.$$

From the Banach lemma on invertible functions [7,14], it follows that $[\lambda_0,\ \delta_0;\ \Omega] \neq 0$ and:

$$\left|[\lambda_0,\ \delta_0;\ \Omega]^{-1}\Omega'(p_*)\right| \leq \frac{1}{1 - \gamma(1 + \alpha)|\lambda_0 - p_*|}. \tag{25}$$

In view of (14) and (18), we have:

$$\left|\Omega'(p_*)^{-1}\Omega(\lambda_0)\right| = \left|\Omega'(p_*)^{-1}\left(\Omega(\lambda_0) - \Omega(p_*)\right)\right|$$
$$= \left|\Omega'(p_*)^{-1}[\lambda_0,\, p_*,\, \Omega](\lambda_0 - p_*)\right| \tag{26}$$
$$\leq M|\lambda_0 - p_*|$$

and similarly:

$$\left|\Omega'(p_*)^{-1}\Omega(\delta_0)\right| \leq M|\delta_0 - p_*|, \tag{27}$$

since $\delta_0 \in \mathbb{P}$. Then, using the second substep of Methods (2), (11), (14), (16), (25) and (27), we obtain:

$$|\mu_0 - p_*| = \left|\delta_0 - p_* - [\lambda_0,\, \delta_0,\, \Omega]^{-1}\Omega(\delta_0)\right|$$
$$\leq \left|[\lambda_0,\, \delta_0,\, \Omega]^{-1}\Omega'(p_*)\right|\left|\Omega'(p_*)^{-1}\left([\lambda_0,\, \delta_0,\, \Omega](\delta_0 - p_*) - (\Omega(\delta_0) - \Omega(p_*))\right)\right|$$
$$\leq \left|[\lambda_0,\, \delta_0,\, \Omega]^{-1}\Omega'(p_*)\right|\left|\Omega'(p_*)^{-1}\left([\lambda_0,\, \delta_0,\, \Omega] - [\delta_0,\, p_*,\, \Omega]\right)(\delta_0 - p_*)\right|$$
$$\leq \frac{\gamma\left(|\lambda_0 - \delta_0| + |\delta_0 - p_*|\right)|\delta_0 - p_*|}{1 - \gamma(1 + \alpha)|\lambda_0 - p_*|} \tag{28}$$
$$\leq \frac{\gamma\left(|\beta|bM|\lambda_0 - p_*| + \alpha|\lambda_0 - p_*|\right)\alpha|\lambda_0 - p_*|}{1 - \gamma(1 + \alpha)|\lambda_0 - p_*|}$$
$$\leq \frac{\gamma\alpha\left(|\beta|bM + \alpha\right)|\lambda_0 - p_*|^2}{1 - \gamma(1 + \alpha)|\lambda_0 - p_*|}$$
$$= g_1(|\lambda_0 - p_*|)|\lambda_0 - p_*| < |\lambda_0 - p_*| < R,$$

and so, (21) is true for $s = 0$ and $\mu_0 \in U(p_*,\, R)$. Next, we need to show that $\Omega(\lambda_0) \neq 0$ and $\Omega(\delta_0) \neq 0$, for $\delta_0 \neq p_*$. Using (14) and (15), and the definition of R, we obtain:

$$\left|\left((\lambda_0 - p_*)\Omega'(p_*)\right)^{-1}\left[\Omega(\lambda_0) - \Omega(p_*) - \Omega'(p_*)(\lambda_0 - p_*)\right]\right|$$
$$\leq |\lambda_0 - p_*|^{-1}\left|\Omega'(p_*)^{-1}\left([\lambda_0,\, p_*;\, \Omega] - \Omega'(p_*)(\lambda_0 - p_*)\right)\right| \tag{29}$$
$$\leq \gamma|\lambda_0 - p_*|^{-1}|\lambda_0 - p_*|^2 = \gamma|\lambda_0 - p_*| < \gamma R < 1.$$

Hence, $\Omega(\lambda_0) \neq 0$ and:

$$|\Omega'(\lambda_0)^{-1}\Omega'(p_*)| \leq \frac{1}{|\lambda_0 - p_*|(1 - \gamma|\lambda_0 - p_*|)}. \tag{30}$$

Similarly, we have that:

$$|\Omega'(\delta_0)^{-1}\Omega'(p_*)| \leq \frac{1}{|\delta_0 - p_*|(1 - \gamma|\delta_0 - p_*|)} \leq \frac{1}{|\delta_0 - p_*|(1 - \alpha\gamma|\lambda_0 - p_*|)}. \tag{31}$$

Then, by using (4) and (12) (for $\delta_0 = \mu_0$), (16), (27), (28), (30) and (31), we have:

$$|H(v_0,\, \eta_0)| \leq |H(|v_0|,\, |\eta_0|)|$$
$$\leq \left|H\left(\frac{M|\mu_0 - p_*|}{|\delta_0 - p_*|(1 - \gamma|\delta_0 - p_*|)},\, \frac{M|\mu_0 - p_*|}{|\lambda_0 - p_*|(1 - \gamma|\lambda_0 - p_*|)}\right)\right|$$
$$\leq \left|H\left(\frac{M\gamma(|\beta|Mb + \alpha)|\lambda_0 - p_*||\delta_0 - p_*|}{|\delta_0 - p_*|(1 - \alpha\gamma|\lambda_0 - p_*|)(1 - \gamma(1 + \alpha)|\lambda_0 - p_*|)},\, \frac{Mg_1(|\lambda_0 - p_*|)|\lambda_0 - p_*|}{|\lambda_0 - p_*|(1 - \gamma|\lambda_0 - p_*|)}\right)\right| \tag{32}$$
$$\leq \left|H\left(\frac{M\gamma(|\beta|Mb + \alpha)|\lambda_0 - p_*|}{(1 - \gamma(1 + \alpha)|\lambda_0 - p_*|)(1 - \alpha\gamma|\lambda_0 - p_*|)},\, \frac{Mg_1(|\lambda_0 - p_*|)}{1 - \gamma|\lambda_0 - p_*|}\right)\right|$$
$$= h(|\lambda_0 - p_*|).$$

Adopting (13), we get:

$$\left| \Omega'(p_*)^{-1} \left([\delta_0, \, \mu_0; \, \Omega] - \Omega'(p_*) \right) \right| \leq \gamma \left(|\delta_0 - p_*| + |\mu_0 - p_*| \right)$$

$$\leq \gamma \left(\alpha |\lambda_0 - p_*| + g_1(|\lambda_0 - p_*|)|\lambda_0 - p_*| \right) \tag{33}$$

$$= q(|\lambda_0 - p_*|) < q(R) < 1.$$

Hence, we have:

$$\left| [\delta_0, \, \mu_0; \, \Omega]^{-1} \Omega'(p_*) \right| \leq \frac{1}{1 - q(|\lambda_0 - p_*|)}. \tag{34}$$

Furthermore, λ_1 is well defined by (24), (32) and (34). Using the third substep of (2), (12), (27) (for $\delta_0 = \mu_0$), (28), (32) and (34), we get:

$$|\lambda_1 - p_*| \leq |\mu_0 - p_*| + |H(v_0, \, \eta_0)| \left| [\lambda_0, \, \delta_0; \, \Omega]^{-1} \Omega'(p_*) \right| \left| \Omega'(p_*)^{-1} \Omega(\mu_0) \right|$$

$$\leq \left[1 + \frac{Mh(|\lambda_0 - p_*|)}{1 - q(|\lambda_0 - p_*|)} \right] |\mu_0 - p_*|$$

$$\leq \left[1 + \frac{Mh(|\lambda_0 - p_*|)}{1 - q(|\lambda_0 - p_*|)} \right] g_1(|\lambda_0 - p_*|)|\lambda_0 - p_*| \tag{35}$$

$$\leq g_2(|\lambda_0 - p_*|)|\lambda_0 - p_*| < |\lambda_0 - p_*| < R,$$

showing that (22) is true for $s = 0$ and $\lambda_1 \in U(p_*, \, R)$. Replacing λ_0, δ_0, and μ_0 by λ_s, δ_s, and μ_s, respectively, in the preceding estimates, we arrive at (20)–(22). From the estimates $\|\delta_{s+1} - p_*\| < \|\delta_s - p_*\| < r$, we conclude that $\lim_{s \to \infty} \delta_s = p_*$ and $x_{s+1} \in U(p_*, \, R)$. Finally, to illustrate the uniqueness, let $p_{**} \in \bar{U}(p_*, T)$ such that $\Omega(p_{**}) = 0$. We assume $Q = [p_*, p_{**}; \Omega]$. Adopting (15), we get:

$$\left| \Omega'(p_*)^{-1} (Q - \Omega'(p_*)) \right| \leq \gamma \left(|p_* - p_*| + |p_{**} - p_*| \right)$$

$$= \gamma T < 1. \tag{36}$$

Therefore, $Q \neq 0$, and in view of the identity $\Omega(p_*) - \Omega(p_{**}) = Q(p_* - p_{**})$, we conclude that $p_* = p_{**}$. $\square$

Remark 1.

(a) *It follows from condition* (15) *and the estimate:*

$$\left| \Omega'(p_*)^{-1}[x, \, p_*; \, \Omega] \right| \quad = \quad \left| \Omega'(p_*)^{-1}([x, \, p_*; \, \Omega] - \Omega'(p_*) - \Omega'(p_*)) + I \right|$$

$$\leq \quad 1 + \left| \Omega'(p_*)^{-1}([x, \, p_*; \, \Omega] - \Omega'(p_*)) \right|$$

$$\leq \quad 1 + \gamma |\lambda_0 - p_*|$$

and Condition (14) *can be discarded and M substituted by:*

$$M = M(v) = 1 + \gamma v$$

or $M = 2$, since $v \in [0, \frac{1}{\gamma})$.

(b)   We note that (2) does not change if we adopt the conditions of Theorem 1 instead of the stronger ones given in [3]. In practice, for the error bounds, we can consider the computational order of convergence (COC) [10]:

$$\xi = \frac{\ln\dfrac{|\delta_{s+2} - p_*|}{|\delta_{s+1} - p_*|}}{\ln\dfrac{|\delta_{s+1} - p_*|}{|\delta_s - p_*|}}, \quad \text{for each } s = 0,1,2,\ldots \tag{37}$$

or the approximated computational order of convergence (ACOC) [10]:

$$\xi^* = \frac{\ln\dfrac{|\delta_{s+2} - \delta_{s+1}|}{|\delta_{s+1} - \delta_s|}}{\ln\dfrac{|\delta_{s+1} - \delta_s|}{|\delta_s - \delta_{s-1}|}}, \quad \text{for each } s = 1,2,\ldots \tag{38}$$

In practice, we obtain the order of convergence that, avoiding the bounds, involves estimates higher than the first Fréchet derivative.

3. Numerical Examples

We consider some of the weight functions to solve a variety of univariate problems that are depicted in Examples 1–3.

Tables 1–3 display the minimum number of iterations necessary to obtain the required accuracy for the zeros of the functions $\Omega(x)$ in Examples 1–3. Moreover, we include also the initial guess, the radius of convergence of the corresponding function, and the theoretical order of convergence. Additionally, we calculate the COC approximated by means of (37) and (38).

All computations used the package *Mathematica* 9 with multiple precision arithmetic, adopting $\epsilon = 10^{-50}$ as a tolerance error and the stopping criteria:

(i) $|\delta_{s+1} - \delta_s| < \epsilon$ and (ii) $|\Omega(\delta_s)| < \epsilon$.

Example 1. *Let* $\mathbb{S} = \mathbb{R}$, $\mathbb{P} = [-\pi,\ \pi], x^* = 0$. *Let us define function* Ω *on* $\mathbb{P}$ *by:*

$$\Omega(x) = \cos x - x - 1. \tag{39}$$

Consequently, it results $\alpha = 1 + \dfrac{|\beta| + M^k|\Omega'(p_*)|^{k-1}}{\gamma^{k-1}}$, $\gamma = \frac{1}{2}$, $b = |\Omega'(p_*)| = 1$ *and* $M = 2$. *We obtain a different radius of convergence when using distinct types of weight functions (for details, please see [3]), COC* (ξ) *and s presented in Table* 1.

Table 1. Radii of convergence according to the adopted weight function.

Cases	Different Values of the Parameters that Satisfy Theorem 1							
	β	k	$H(v, \eta)$	R_1	R_q	R	λ_0	s	ξ
1.	-1	1	$\frac{1+v}{1-\eta}$	0.10526	0.27008	0.02535	0.024	4	4
2.	3	2	$1+2v$	0.00250	0.03749	0.00082	0.0007	3	4
3.	-3	3	$1+2\eta$	0.00020	0.01013	0.00004	0.0003	3	4
4.	0.1	4	$\frac{1}{1-2\eta}$	0.00962	0.07090	0.00160	0.0005	3	4

Example 2. *Let* $\mathbb{S} = \mathbb{R}, \mathbb{P} = [-1,\ 1], x^* = 0.714806$ *(approximated root), and let us assume function* Ω *on* $\mathbb{P}$ *by*

$$\Omega(x) = e^x - 4x^2. \tag{40}$$

As a consequence, we get $\alpha = 1 + \frac{|\beta| + M^k |\Omega'(p_*)|^{k-1}}{\gamma^{k-1}}$, $\gamma = 2$, $b = |\Omega'(p_*)| = |e^{x^*} - 8p_*| \approx 3.67466$ *and* $M = 2$. *We have the distinct radius of convergence when using several weight functions (for details, please see* [3]*), COC* (ξ) *and s listed in Table* 2.

Table 2. Radii of convergence according to the adopted weight function.

Cases	Different Values of the Parameters that Satisfy Theorem 1								
	β	k	$H(v, \eta)$	R_1	R_q	R	λ_0	s	ξ
1.	-1	1	$\frac{1+v}{1-\eta}$	0.01427	0.05498	0.00318	0.713	4	4
2.	3	2	$1 + 2v$	0.00047	0.00896	0.00015	0.7417	4	4
3.	-3	3	$1 + 2\eta$	0.00006	0.00286	0.00001	0.7418	3	4
4.	0.1	4	$\frac{1}{1-2\eta}$	0.00359	0.02201	0.00060	0.7413	4	4

Example 3. *Using the example of the introduction, we have* $\alpha = 1 + \frac{|\beta| + M^k |\Omega'(p_*)|^{k-1}}{\gamma^{k-1}}$, $\gamma = 2$, $b = |\Omega'(p_*)| = \frac{2\pi + 1}{\pi^3} \approx 0.23489$, $M = 2$, *and the required zero is* $p_* = \frac{1}{\pi} \approx 0.318309886$. *We have different radii of convergence by adopting distinct types of weight functions (for details, please see* [3]*), COC* (ξ) *and s in Table* 3.

Table 3. Radii of convergence according to the adopted weight function.

Cases	Different Values of the Parameters that Satisfy Theorem 1								
	β	k	$H(v, \eta)$	R_1	R_q	R	λ_0	s	ξ
1.	-1	1	$\frac{1+v}{1-\eta}$	0.03470	0.07391	0.00884	0.325	4	4
2.	3	2	$1 + 2v$	0.03965	0.08356	0.01225	0.329	4	4
3.	-3	3	$1 + 2\eta$	0.08363	0.13140	0.02437	0.298	5	4
4.	0.1	4	$\frac{1}{1-2\eta}$	0.16367	0.18912	0.05268	0.358	5	4

4. Conclusions

Locating the range or interval of the required root that provides sure convergence of an iterative method is one of the difficult problems in computational analysis. This paper addressed this problem and expanded the applicability of Methods (2) and (3) using hypotheses only on the functions appearing in these techniques. Further, we provided the radii of ball convergence and error bounds using Lipschitz conditions. This type of study was not addressed in the earlier work. With the help of the radius of convergence, we can find the range of initial estimate p_* that tells us how close it must be for granting the convergence of Methods (2) and (3). Finally, the applicability of new approach was illustrated with several numerical examples.

Author Contributions: All co-authors contributed to the conceptualization, methodology, validation, formal analysis, writing the original draft preparation, and editing.

Funding: This research received no external funding.

Conflicts of Interest: The authors declare no conflict of interest.

References

1. Traub, J.F. *Iterative Methods for the Solution of Equations*; Prentice-Hall Series in Automatic Computation; Prentice-Hall: Englewood Cliffs, NJ, USA, 1964.

2. Petkovic, M.S.; Neta, B.; Petkovic, L.; Džunič, J. *Multipoint Methods For Solving Nonlinear Equations*; Elsevier: Amsterdam, The Netherlands, 2013.

3. Lee, M.Y.; Kim, Y.I. A family of fast derivative-free fourth order multipoint optimal methods for nonlinear equations. *Int. J. Comput. Math.* **2012**, *89*, 2081–2093. [CrossRef]

4. Amat, S.; Busquier, S.; Plaza, S. Dynamics of the King and Jarratt iterations. *Aequ. Math.* **2005**, *69*, 212–223. [CrossRef]

5. Amat, S.; Busquier, S.; Plaza, S. Chaotic dynamics of a third-order Newton-type method. *J. Math. Anal. Appl.* **2010**, *366*, 24–32. [CrossRef]

6. Amat, S.; Hernández, M.A.; Romero, N. A modified Chebyshev's iterative method with at least sixth order of convergence. *Appl. Math. Comput.* **2008**, *206*, 164–174. [CrossRef]

7. Argyros, I.K. *Convergence and Application of Newton-Type Iterations*; Springer: Berlin/Heidelberg, Germany, 2008.

8. Argyros, I.K.; Hilout, S. *Numerical Methods in Nonlinear Analysis*; World Scientific Publ. Comp: River Edge, NJ, USA, 2013.

9. Behl, R.; Motsa, S.S. Geometric construction of eighth-order optimal families of Ostrowski's method. *Recent Theor. Appl. Approx. Theory* **2015**, *2015*, 614612. [CrossRef] [PubMed]

10. Ezquerro, J.A.; Hernández, M.A. New iterations of R-order four with reduced computational cost. *BIT Numer. Math.* **2009**, *49*, 325–342. [CrossRef]

11. Kanwar, V.; Behl, R.; Sharma, K.K. Simply constructed family of a Ostrowski's method with optimal order of convergence. *Comput. Math. Appl.* **2011**, *62*, 4021–4027. [CrossRef]

12. Magreñán, Á.A. Different anomalies in a Jarratt family of iterative root-finding methods. *Appl. Math. Comput.* **2014**, *233*, 29–38.

13. Magreñán, Á.A. A new tool to study real dynamics: The convergence plane. *Appl. Math. Comput.* **2014**, *248*, 215–224. [CrossRef]

14. Rheinboldt, W.C. An adaptive continuation process for solving systems of nonlinear equations. *Pol. Acad. Sci. Banach Cent. Publ.* **1978**, *3*, 129–142. [CrossRef]

15. Weerakoon, S.; Fernando, T.G.I. A variant of Newton's method with accelerated third order convergence. *Appl. Math. Lett.* **2000**, *13*, 87–93. [CrossRef]

Article

Extending the Adapted PageRank Algorithm Centrality to Multiplex Networks with Data Using the PageRank Two-Layer Approach

Taras Agryzkov [1], Manuel Curado [2], Francisco Pedroche [3], Leandro Tortosa [1] and José F. Vicent [1,*]

[1] Campus de San Vicente, Department of Computer Science and Artificial Intelligence, University of Alicante, Ap. Correos 99, E-03080 Alicante, Spain; taras.agryzkov@ua.es (T.A.); tortosa@ua.es (L.T.)
[2] Campus de los Canteros, Departament of Technology, Catholic University of Ávila, Los Canteros s/n, E-05005 Ávila, Spain; manuel.curado@ucavila.es
[3] Institut de Matemàtica Multidisciplinària, Universitat Politècnica de València, Camí de Vera s/n, E-46022 València, Spain; pedroche@mat.upv.es
* Correspondence: jvicent@ua.es; Tel.: +34-965-903900

Received: 25 January 2019; Accepted: 19 February 2019; Published: 22 February 2019

Abstract: Usually, the nodes' interactions in many complex networks need a more accurate mapping than simple links. For instance, in social networks, it may be possible to consider different relationships between people. This implies the use of different layers where the nodes are preserved and the relationships are diverse, that is, multiplex networks or biplex networks, for two layers. One major issue in complex networks is the centrality, which aims to classify the most relevant elements in a given system. One of these classic measures of centrality is based on the PageRank classification vector used initially in the Google search engine to order web pages. The PageRank model may be understood as a two-layer network where one layer represents the topology of the network and the other layer is related to teleportation between the nodes. This approach may be extended to define a centrality index for multiplex networks based on the PageRank vector concept. On the other hand, the adapted PageRank algorithm (APA) centrality constitutes a model to obtain the importance of the nodes in a spatial network with the presence of data (both real and virtual). Following the idea of the two-layer approach for PageRank centrality, we can consider the APA centrality under the perspective of a two-layer network where, on the one hand, we keep maintaining the layer of the topological connections of the nodes and, on the other hand, we consider a *data layer* associated with the network. Following a similar reasoning, we are able to extend the APA model to spatial networks with different layers. The aim of this paper is to propose a centrality measure for biplex networks that extends the adapted PageRank algorithm centrality for spatial networks with data to the PageRank two-layer approach. Finally, we show an example where the ability to analyze data referring to a group of people from different aspects and using different sets of independent data are revealed.

Keywords: adapted PageRank algorithm; PageRank vector; networks centrality; multiplex networks; biplex networks

1. Introduction

1.1. Literature Review

Network theory is an important tool for describing and analyzing complex systems throughout the social, biological, physical, information and engineering sciences [1]. Originally, almost all studies of networks employed an abstraction in which systems are represented as ordinary graphs [2]. Although this approach is naive in many respects, it has been extremely successful. For instance, it has been used to illustrate that many real networks possess a heavy-tailed degree distribution [3], exhibit the small-world property [4], contain nodes that play central roles [5] and/or have modular structures [6].

It has been recently recognized [7–9] that most complex systems are not simply formed by a simple network, but they are instead formed by multilayer networks. Multilayer networks include not just one but several layers (networks) characterizing interactions of different nature and connotation. Multiplex networks [10,11] are a special type of multilayer networks. They are formed by a set of nodes connected by different types of interactions. The term multiplex was defined to indicate the presence of more than one relationship between the same actors of a social network [12]. Each set of interactions of the same type determines a distinct layer (network) of the multiplex. Examples of multiplex networks are ubiquitous. Other major examples of multiplex networks range from transportation networks [13] to social [14], financial [15] and biological networks [16]. In transportation networks, different layers can represent different means of transportation while in scientific collaboration networks the different layers can represent several topics of the collaboration. Given the surge of interest in multiplex networks, recently several algorithms [17,18] have been proposed to assess the centrality of nodes in these multilayer structures.

On the other hand, when pairs of nodes can be connected through multiple links and in multiple layers, the ranking of nodes should necessarily reflect the importance of nodes in one layer as well as their importance in other interdependent layers. Our world is increasingly dependent on efficient ranking algorithms [19–21]. Currently the ranking of nodes in complex networks is used in a variety of different contexts [22], from finance to social, urban and biological networks. For instance, in the context of economical trade networks, formed by networks of countries and products, ranking algorithms [23] are recognized as an important tool to evaluate the economic development of countries.

Within the general problem of centrality in urban networks [24], the adapted PageRank algorithm (APA) proposed by [25] provides us a model to establish a ranking of nodes in spatial networks according to their importance in it. This centrality was originally proposed for urban networks, although it may be generalized to spatial networks or networks with data. It constitutes a centrality measure in urban networks with the main characteristic that it is able to consider the importance of data obtained from any source in the whole process of computing the centrality of the individual nodes. Starting from the basic idea of the PageRank vector concept [26], the matrix used for obtaining the classification of the nodes is constructed in a slightly different way as we will see later.

Centrality measures originally defined on single networks have been used extensively in some types of networks, as social, technological or biological. In multiplex networks these measures can be extended in different ways [27]. Generalizing centrality measures from monoplex networks to multilayer networks is not trivial. When ranking nodes in a multilayer or multiplex network, the key question to be addressed is how one should take into consideration all the different types of edges, not all of which have the same importance [17].

The Multiplex PageRank [28] evaluates the centrality of the nodes of multiplex networks and is based on the idea of biased random walks to define the Multiplex PageRank centrality measure in which the effects of the interplay between networks on the centrality of nodes are directly taken into account. Other centrality measures associate a different influence with the links of different layers that weights their contribution to the centrality of the nodes, as for example Multiplex Eigenvector Centralities [17], or Functional Multiplex PageRank [18], which takes into account the fact that different multilinks contribute differently to the centrality of each node and associates with each node a function. In [29], Solé–Ribalta et al. re-define the betweenness centrality measure to account for the inherent structure of multiplex networks and propose an algorithm to calculate it efficiently.

1.2. Main Contribution

The main objective of this paper is to propose a centrality measure for biplex networks that adapts the APA centrality for spatial networks with data to the PageRank two-layer approach.

The multiplex PageRank algorithm proposed by Halu et al. [28] measures the centrality of a node in a layer β with the corresponding adjacency matrix using a random walker as in the usual PageRank biased by the PageRank of the nodes in the layer *alpha*. So, Multiplex PageRank can be described in terms of the bias that one layer exerts on the random jumps that a surfer makes in another layer. Four versions of the centrality are presented depending on how layers affect each other or, alternatively, exert a bias upon the random walk.

The advantage of the model proposed for biplex networks is that since it is based on the APA centrality, the randomness in the jumps of the surfer is replaced by the influence of the data present in the network. Therefore, the model is able to study and analyze several relationships of a set of nodes by different layers, but it is also capable of measuring the influence of the data in the different layers of the network. Thus, it is not only taken into account the topology of the networks but also the data. In addition, the α parameter initially associated to the PageRank model and exported to APA centrality can be used to assign the importance that each of these data layers has in the calculation of centrality. These characteristics summarizes the novelty and potential of this research work.

The potential of the applications to which this centrality can be applied should also be highlighted. Nowadays multilayers and multiplex networks are investigated in many fields, as were enumerated in Section 1.1. However, there are other potential applications not so well known but equally interesting where the centrality measure proposed in this research can be applied. One of these topics is the improvement in the recommender systems [30,31]. These systems were initially based on demographic, content-based and collaborative filtering although actually are incorporating social information. Another unknown topic of application is in Public Safety Networks [32,33] which have emerged as the key solution to a successful response to emergency and disaster events. In these systems each user is associated with some social, physical and mobility-related characteristics and attributes in a public safety network.

1.3. Structure of the Paper

This paper is organized as follows: Section 2 summarizes the methodology used in the paper: the basic characteristics of the APA centrality, as well as the biplex approach to the PageRank vector. Once these models are described, the new centrality is presented combining the APA centrality (based on data) and the biplex approach of PageRank vector. In Section 3, some results are presented with their interpretation. The discussion of the experimental results are performed in Section 4 and, finally, some conclusions are summarized.

2. Methodology

In practical applications, the computation of a global index to measure the importance to each node is a main task. If the system studied contains several types of relations between actors it is expected that the measures, in some way, consider the importance obtained from the different layers. A simple choice could be to combine the centrality of the nodes computed from the different layers independently according to some heuristic choice.

In this section, both the well-established methods and the proposed centrality for multiplex networks with data are described in detail.

2.1. The Adapted PageRank Algorithm (APA) Model

Let us establish some notation that will be used in the following. Let $\mathcal{G} = (\mathcal{N}, \mathcal{E})$ be a graph where $\mathcal{N} = \{1, 2, \dots, n\}$ and $n \in \mathrm{N}$. The link (i, j) belongs to the set $\mathcal{E}$ if and only if there exists a link connecting node i to j. The adjacency matrix of $\mathcal{G}$ is an $n \times n$ matrix $A = (a_{ij})$, where

$$a_{ij} = \begin{cases} 1 & \text{if } (i,j) \text{ is a link,} \\ 0 & \text{otherwise.} \end{cases}$$

The adapted PageRank algorithm (APA) proposed by Agryzkov et al. [25] provides us a model to establish a ranking of nodes in an urban network taking into account the data presented in it. This centrality was originally proposed for urban networks, although it may be generalized to spatial networks or networks with data. It constitutes a centrality measure for networks with the main characteristic being that it is able to consider the importance of data obtained from any source in the whole process of computing the centrality of the individual nodes. Starting from the basic idea of the PageRank vector concept, the construction of the matrix used for obtaining the classification of the nodes is modified.

In its original approach, PageRank was based on a model of a web surfer that probabilistically browses the web graph, starting at a node chosen at random according to a personalization vector whose components give us the probability of starting at node v. At each step, if the current node had outgoing links to other nodes, the surfer next browsed with probability α one of those nodes (chosen uniformly at random), and with probability $1 - \alpha$ a node chosen at random according to the personalized vector. For the web graph, the most popular value of the dumping factor was 0.85. If the current node was a sink with no outgoing links, the surfer automatically chose the next node at random according to the personalized vector.

In the APA model, the data matrix was constructed following a similar reasoning from the original idea of the PageRank vector; a random walker can jump between connecting nodes following the local link given by the network or can jump between nodes (not directly connected) with the same probability, regardless the topological distance between them (number of nodes in the walk).

In the algorithm implemented to calculate the APA centrality (see [25], p. 2190), a new matrix $A^* = (p_{ij}) \in \mathcal{R}^{n \times n}$ is constructed from the adjacency matrix A, as

$$
p_{ij} = \begin{cases} \frac{1}{c_j} & \text{if } a_{ij} \neq 0, \\ 0 & \text{otherwise,} \end{cases} \quad 1 \leq i, j \leq n, \tag{1}
$$

where c_j represents the sum of the j-th column of the adjacency matrix.

Algebraically, A^* may be obtained as

$$
A^* = A\Delta^{-1}, \tag{2}
$$

where $\Delta = (\delta_{ij}) \in \mathcal{R}^{n \times n}$ is the degree matrix of the graph, that is, $\delta_{ij} = c_j^{-1}$, for $i = j$ and $\delta_{ij} = 0$, for $i \neq j$. We refer to A^* as the transition matrix, and it represents, by columns, the probability to navigate from a page to other. In the literature related to this topic the matrix A^* is also denoted as P or P_A, so A^*, P or P_A are the same matrix. Following the notation of Pedroche et al. [34] it will preferably be used P.

The transition matrix, $P = A^*$ has the following characteristics (see [25]):

1. It is nonnegative.
2. It is stochastic by columns.
3. The highest eigenvalue of P is $\lambda = 1$.

The key point of the model is the construction of the so-called data matrix D of size $n \times k$, with its n rows representing the n nodes of the network, and each of its k columns representing the attributes of the data object of the study. Specifically, an element $d_{ij} \in D$ is the value we attach to the data class k_j at node i.

However, not all the characteristics of data may have the same relevance or influence in the question object of the analysis. Therefore, a vector $v_0 \in \mathbb{R}^{k \times 1}$ is constructed, where the element that occupies the row i is the multiplicative factor associated with the property or characteristic k_j. With this vector v_0 a weighting factor of the data is introduced, in order to work with the entire data set or a part of it.

Then, multiplying D and v_0, v may be obtained as

$$v = D \cdot v_0,$$

with $v \in \mathbb{R}^{n \times 1}$.

The construction of vector v allows us to associate to each node a value that represents the amount of data assigned to it. Thus, two different values are associated with every node; on the one hand, its degree, related to the topology and, on the other hand, the value of the data associated to it. For a more detailed description of how the data are associated to the nodes, see [25,35].

After normalizing v, denoted as v^*, it is possible to define the matrix M_{APA} as

$$M_{APA} = (1 - \alpha)P + \alpha V, \tag{3}$$

where $V \in \mathbb{R}^{n \times n}$ is a matrix in which all of its components in the i-th row are equal to v_i^*. The parameter α is fixed and it is related to the teleportation idea. The value that is traditionally used is $\alpha = 0.15$.

In practice, vector v^* is repeated (n times) in every column of the matrix V.

The matrix M_{APA} was used to compute the ranking vector for the network.

With these considerations, the APA algorithm proposed in [25] may be summarized by the Algorithm 1:

Algorithm 1: (Adapted PageRank algorithm (APA)). *Let $G = (V, E)$ be a primary graph representing a network with n nodes.*

---

1   *Compute the matrix P from the graph G.*
2   *Construct the data matrix D.*
3   *Construct the weighted vector v_0.*
4   *Compute v as $Dv_0 = v$.*
5   *Normalize v, and denote it as v^*.*
6   *Construct V as $V = v^* e^T$.*
7   *Construct the matrix M_{APA} as $M_{APA} = (1 - \alpha) P + \alpha V$.*
8   *Compute the eigenvector x of the matrix M_{APA} associated to eigenvalue $\lambda = 1$. The components of the resulting eigenvector x represent the ranking of the nodes in the graph G.*

The main feature of this algorithm is the construction of the data matrix D and the weighted vector v_0. The matrix D allows us to represent numerically the dataset. Vector v_0 determines the importance of each of the factors or characteristics that have been measured by means of D.

The Perron–Frobenius theorem is of great importance in this problem, since it constitutes the theoretical base that ensures that there exists an eigenvector x associated with the dominant eigenvalue $\lambda = 1$, so that all its components are positive, which allows establishing an order or classification of these elements. In our case, due to the way in which P and V have been constructed, it can be seen that M_{APA} is a stochastic matrix by columns, which assures us of the spectral properties necessary for the Perron–Frobenius theorem to be fulfilled. Therefore, the existence and uniqueness of a dominant eigenvector with all its components positive is guaranteed. See [36,37] for further study of spectral and algebraic properties of the models based on PageRank.

Vector x constitutes the adapted PageRank vector and provides a classification or ranking of the pages according to the connectivity criterion between them and the presence of data.

2.2. The Biplex Approach for Classic PageRank

Pedroche et al. [34] propose a two-layer approach for the classic PageRank classification vector based on the idea that we now briefly expose. The two-layer approach is to consider the PageRank classification of the nodes as a process divided into two clearly differentiated parts. The first part is related to the topology of the network, where the connections of the nodes are basically taken into account by means of their adjacency matrix. There is a second part regarding to the teleportation from one node to another, following a criterion of equiprobability.

They affirm that the PageRank classification for a graph G with personalized vector v can be understood as the stationary distribution of a Markov chain that occurs in a network with two layers, which are:

l_1, **physical layer** , it is the network G.
l_2, **teleportation layer** , it is an all-to-all network, with weights given by the personalized vector.

Under this perspective, it is easy to construct a block matrix M_A based on these two layers where each of the diagonal blocks is associated to a given layer. Therefore, we can construct

$$M_A = \left(\begin{array}{c|c} \alpha P_A & (1-\alpha)I \\ \hline 2\alpha I & (1-\alpha)ev^T \end{array} \right) \quad \in \mathbb{R}^{2n \times 2n}, \tag{4}$$

where M_A defines a Markov chain in a network with two layers.

Due to the good spectral characteristics of M_A (it is irreducible and primitive), they arrive to the conclusion that given a network with n nodes, whose adjacency matrix is A, the two-layer approach PageRank of A is the vector

$$\hat{\pi}_A = \pi_u + \pi_d \in \mathbb{R}^n,$$

where $\left[\pi_u^T \ \pi_d^T\right]^T \in \mathbb{R}^{2n}$ is the unique normalized and positive eigenvector of matrix M_A given by the expression (4).

In [38], the authors propose a new centrality measure for complex networks with geo-located data based on the application of the two-layer PageRank approach to the APA centrality measure for spatial networks with data. They design an algorithm to evaluate this centrality and show the coherence of this measure regarding the original APA by calculating the correlation and the quantitative difference of both centralities using different network models. This coherence in the results obtained for the APA and the proposed centrality using the two-layer approach is absolutely mandatory in our objective to extend the two-layer approach for multiplex networks with data.

Therefore, the two-layer approach may be extended to the case of multiplex networks, where we have several networks with the same nodes and with different topologies and connections between nodes. Following the notation used by Pedroche et al. [34], let us consider a multiplex network $\mathcal{M} = (\mathcal{N}, \mathcal{E}, \mathcal{S})$ with layers $\mathcal{S} = (l_1, l_2, \ldots, l_k)$. Given a multiplex network $\mathcal{M}$ with several layers, a multiplex PageRank centrality can be defined by associating to each layer l_i a two-layer random walker with one the physical layer and a teleportation layer. In addition, transition between these layers must be allowed. The idea behind this process is the application of the two-layer approach to each layer of the multiplex network.

For now, let us consider our problem restricted to biplex networks $\mathcal{M} = (\mathcal{N}, \mathcal{E}, \mathcal{S})$, with layers $\mathcal{S} = (l_1, l_2)$ whose adjacency matrices are $A_1, A_2 \in R^{n \times n}$, respectively. For convenience in the notation we will write P_A, P_1 and P_2 instead of A^*, A_1^* and A_2^*, respectively.

The authors (see, [34]) construct a general matrix M_2 as a new block matrix by associating to each layer l_i a two-layer multiplex defined, for $i = 1, 2$, as:

$$M_{i,i} = \left(\begin{array}{c|c} \alpha P_A & (1-\alpha)I \\ \hline 2\alpha I & (1-\alpha)ev^T \end{array} \right) \quad \in \mathbb{R}^{2n \times 2n}.$$

Reordering the blocks in such a way that the physical layers appear in the first block, the final matrix is

$$
M_2 = \frac{1}{2}
\left(
\begin{array}{cc|cc}
\alpha P_1 & I & (1-\alpha)I & 0 \\
I & \alpha P_2 & 0 & (1-\alpha)I \\
2\alpha I & 0 & (1-\alpha)ev_1^T & (1-\alpha)ev_2^T \\
0 & 2\alpha I & (1-\alpha)ev_1^T & (1-\alpha)ev_2^T,
\end{array}
\right)
\tag{5}
$$

with P_i, for $i = 1,2$ row stochastic matrices and v_i, for $i = 1,2$ the personalized vectors.

It is straightforward to check that all the spectral properties of M_2 are essentially the same as the Google matrix in the PageRank model. Then, there exists an eigenvector

$$
\hat{\pi}_2 = (\pi_{u_1}, \pi_{u_2}, \pi_{d_1}, \pi_{d_2}) \in \mathbb{R}^{4n}
$$

associated to the dominant eigenvalue $\lambda = 1$. This vector is the key to obtain the classification vector representing the nodes centrality.

Consequently, summarizing the main characteristic of the biplex PageRank approach, considering a biplex network $\mathcal{M}$ with n nodes, with two layers $\mathcal{S} = (l_1, l_2)$, and whose adjacency matrices are $A_1, A_2 \in \mathbb{R}^n$, it can be affirmed that the PageRank vector that classifies the nodes of this biplex network is the unique vector π_2 such as

$$
\pi_2 = \frac{1}{2}\left(\pi_{u_1} + \pi_{u_2} + \pi_{d_1} + \pi_{d_2}\right) \in \mathbb{R}^n,
$$

with π_2 normalized.

2.3. Constructing the APABI Centrality by Applying the Two-Layer Approach

The idea of the treatment of the PageRank concept by means of two layers has a great sense within the idea of APA centrality, since the influence of the data in the network is measured separately in the original algorithm. Paying attention to the construction of M_{APA} given by (3), note that V is the matrix summarizing all the data information. But not only the application of this concept is interesting for our centrality, since may be also interesting to analyze the differences that occur between both techniques of calculating the importance of the nodes.

In this section, we describe how to adapt the APA centrality taking as a reference the two-layers technique, where a block matrix is used to distinguish the topology and the personalized vector.

The original APA centrality model, described in Section 2, presents some differences from the model described and implemented by Pedroche et al. [34], where the final matrix involved in the eigenvector computation is stochastic by rows. In our approach, the basis of the original APA model consists of the construction of a stochastic matrix by columns, where we reflect the topology of the network by the probability matrix P and the influence of the data, through the matrix V.

In order to build a 2×2 block matrix, the same approach used in [34] may be reproduced. The first upper diagonal block contains the information referring to the network topology, while the lower diagonal block is associated to the collected data in the network and assigned to each node of it.

Taking as a reference the APA algorithm, the matrix used to compute the eigenvector associated to the dominant eigenvalue $\lambda = 1$ is given by

$$
M_{APA} = (1-\alpha)P + \alpha V.
$$

A new 2×2 block matrix is constructed as

$$
M_{APA2} =
\left(
\begin{array}{c|c}
\alpha P_A & (1-\alpha)I \\
\hline
\alpha I & \alpha V
\end{array}
\right)
\in \mathbb{R}^{2n \times 2n}.
\tag{6}
$$

The idea that underlies the construction of the matrix by blocks given by (6) is to maintain the spectral properties of the original matrix M_{APA}, with the aim that the numerical algorithms for determining dominant eigenvalue and eigenvector are stable and fast. Note that we have doubled the size of the original matrix.

Following the same reasoning used in Section 3 to construct a model for biplex networks taking as a basis the classical PageRank vector, it is necessary to extend the two-layers APA approach given by the block matrix (6). Using the same notation, let us consider $\mathcal{M} = (\mathcal{N}, \mathcal{E}, \mathcal{S})$, with layers $\mathcal{S} = (l_1, l_2)$ be a biplex network.

Reordering the blocks in such a way that the physical layers appear in the first block, the final matrix is given by

$$M_{BI} = \frac{1}{2} \left(\begin{array}{cc|cc} (1-\alpha)P_1 & I & 2(1-\alpha)I & 0 \\ I & (1-\alpha)P_2 & 0 & 2(1-\alpha)I \\ \hline I & 0 & \alpha V_1 & \alpha V_2 \\ 0 & I & \alpha V_1 & \alpha V_2, \end{array} \right) \tag{7}$$

with P_i, for $i = 1, 2$ column stochastic matrices and V_i, for $i = 1, 2$, the matrices containing the data information.

Note the differences between matrices M_2 (5) and M_{BI} (7). The matrix M_2 is stochastic by rows, however, in the APA centrality the basic matrix M_{APA} is stochastic by columns, so the definition of the matrix M_{BI} is determined by the need to maintain the spectral properties suitable for obtaining the proper vector of the centrality. These desirable spectral properties are ensured by the way in which M_{BI} has been built, being stochastic by columns, as well as irreducible.

Then, there exists an eigenvector

$$\hat{\pi}_{BI} = (\pi_{u_1}, \pi_{u_2}, \pi_{d_1}, \pi_{d_2}) \in \mathbb{R}^{4n} \tag{8}$$

associated to the dominant eigenvalue $\lambda = 1$. This vector is the key to obtain the classification vector representing the nodes centrality. Therefore, it can be obtained a unique vector

$$x = \frac{1}{2} (\pi_{u_1} + \pi_{u_2} + \pi_{d_1} + \pi_{d_2}) \in \mathbb{R}^n, \tag{9}$$

with x a normalized vector.

In Figure 1, a schematic representation of the extended APA model for biplex networks has been represented. All the graphs share the same n nodes, although the relationships between them in the two layers l_1 and l_2 are different, which produces two different adjacency matrices A_1 and A_2. Data are also different in each layer; consequently, two data matrices are constructed D_1 and D_2.

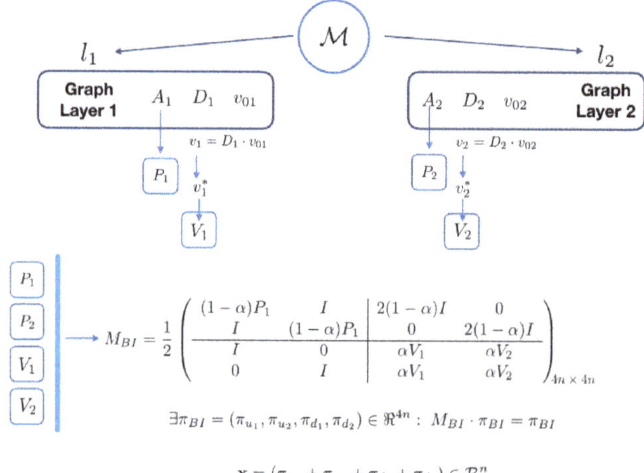

Figure 1. Scheme of the adapted PageRank algorithm (APA) extended model for biplex networks.

The existence of the weighted vectors v_{01} and v_{02} allows us to determine those data that are the object of our interest, being able to discard those that do not present real interest for the study that is carried out.

The model presented in this section may be summarized by the Algorithm 2.

Algorithm 2: (Adapted PageRank algorithm biplex (APABI)). *Let $\mathcal{M} = (\mathcal{N}, \mathcal{E}, \mathcal{S})$, with layers $\mathcal{S} = (l_1, l_2)$ and adjacency matrices A_1, A_2 be a biplex network with n nodes.*

---

1 *For the layers l_i, with $i = 1, 2$, construct the probability matrices P_i, for $i = 1, 2$, respectively.*
2 *From data, construct the matrices D_i, for $i = 1, 2$, respectively.*
3 *Define the weighted vectors v_i, for each layer.*
4 *For the layers l_i, for $i = 1, 2$, compute v_i as $D_i v_{0i}$, respectively.*
5 *Normalize v_i, for $i = 1, 2$, and denote them as v_i^*.*
6 *For the layers l_i, for $i = 1, 2$, construct V_i as $V_i = v_i^* e^T$, respectively.*
7 *From P_i, V_i, for $i = 1, 2$, and the parameter α, construct M_{BI} using the expression (7).*
8 *Compute the eigenvector π_{BI} using the expression (8).*
9 *The components of the resulting eigenvector x, given by the expression (9), represent the ranking of the nodes in the biplex network.*

Algorithm 2 summarizes the steps necessary to calculate a centrality measure that will be denoted as the adapted PageRank algorithm biplex (APABI). This measure provides us with a vector for classifying the nodes of the network according to their importance within a biplex network. This classification is obtained from the importance of the nodes in two networks where what changes are the associations between the nodes and the data associated with them. This classification is obtained from the importance of the nodes in two layers where the nodes are the same and what changes are the associations (links) between the nodes and the data associated with them.

Note that the M_{BI} matrix is built for biplex networks. However, it can be easily extended the same reasoning for a multiplex network with k layers $\{l_1, l_2, \ldots, l_k\}$, defining the adjacency matrices $\{A_1, A_2, \ldots Al_k\}$ and a set of k data matrices $\{D_1, D_2, \ldots, D_k\}$. The matrix M_{BI} is extended to a multiplex network with k layers as

$$M_{multi} = \frac{1}{k} \left(\begin{array}{c|c} M_{1,1} & M_{1,2} \\ \hline M_{2,1} & M_{2,2} \end{array} \right)$$

with

$$M_{1,1} = \begin{pmatrix} (1-\alpha)P_1 & I & \cdots & I \\ I & (1-\alpha)P_2 & \cdots & I \\ \cdots & \cdots & \cdots & \cdots \\ I & I & \cdots & (1-\alpha)P_k \end{pmatrix}, \tag{10}$$

$$M_{2,2} = \begin{pmatrix} \alpha V_1 & \alpha V_2 & \cdots & \alpha V_k \\ \alpha V_1 & \alpha V_2 & \cdots & \alpha V_k \\ \cdots & \cdots & \cdots & \cdots \\ \alpha V_1 & \alpha V_2 & \cdots & \alpha V_k \end{pmatrix}, \tag{11}$$

where $M_{1,2}$, $M_{2,1}$ are diagonal matrices. More exactly, $M_{1,2}$ is formed by k blocks $2(1-\alpha)I$ and $M_{2,1}$ is formed by k identity blocks I.

In the approach made so far in this section, we have considered the same value for the parameter α in all the layers that make up the network. However, it could happen that the α value was different in the different layers, as a consequence of the need to differentiate the importance that must be assigned to the data in each of the layers. That leads us to consider various α_i, for each layer i. Note that this variant does not imply in any way modify the spectral properties of the matrices involved in the calculation of centrality. Consequently, matrix M_{BI} should be written now as

$$M_{BI} = \frac{1}{2} \left(\begin{array}{cc|cc} (1-\alpha_1)P_1 & I & 2(1-\alpha_1)I & 0 \\ I & (1-\alpha_2)P_2 & 0 & 2(1-\alpha_2)I \\ \hline I & 0 & \alpha_1 V_1 & \alpha_2 V_2 \\ 0 & I & \alpha_1 V_1 & \alpha_2 V_2 \end{array} \right) \qquad (12)$$

The generalization of the matrix M_{BI} to k layers with k parameters α_i consists simply of replacing each α in the expressions (10) and (11) with its corresponding α_i in the i-th row.

2.4. A Note About the Computational Cost

We discuss certain general aspects of the computational cost of the proposed model. It should be noted that if we look closely at Algorithm 2, the most expensive algebraic operations that are carried out are the product of a scalar by a matrix, the matrix-vector product and the calculation of the dominant eigenpair $(\lambda, \mathbf{x})$ of matrix M_{BI}, given by the expression (8).

As it is well-known and can be seen in any linear algebra textbook, the product of a scalar by a square matrix of size n requires $n \times n$ multiplications, while the product of two square matrices of size n requires a computational cost of $O(n^2)$. In our case, we need to make the product $D \cdot v_i$, for $i = 1, 2$, where D is the data matrix of size $n \times k$ and v_i is a column vector of size n. Therefore, the computational cost of $D \cdot v_0$ is of $O(nk)$.

However, the most expensive part from the computational point of view is found in step 8 of the algorithm, in which, once the M_{BI} matrix is constructed, it is necessary to obtain its dominant eigenpair $(\lambda, \mathbf{x})$. The numerical problem of calculating the eigenvalues and eigenvectors of any matrix is very expensive, in general, if the matrix in question does not have a structure that simplifies its calculation in some way. In general, for matrices of low dimension (such as $N < 150$), there are efficient methods for finding all the eigenvalues and eigenvectors. For example, the Householder–QL–Wilkinson modification of the given method is built into the EISPACK routines and is routinely used. The computation time for any of these methods grows as N^3 and the memory requirement grows as N^2. For large matrices, a very commonly used algorithm is Lanczos, that is an adaptation of power methods to find the m most useful eigenvalues and eigenvectors of an $n \times n$ Hermitian matrix. For a more detailed description of the numerical matrix eigenvalue problems, see [39].

Due to the way we have built the M_{BI} matrix following the original idea of the Google matrix used in the original PageRank combined with the two-layer PageRank approach, we have ensured that this matrix inherits the spectral properties of the original Google matrix in the original PageRank model. It is a stochastic matrix by columns of which we can affirm, using a variant of the Perron–Frobenius theorem, that its own dominant eigenvector associated to eigenvalue $\lambda = 1$ corresponds to the stationary distribution of the Markov chain by the column normalized matrix M_{BI}. This stationary vector $\hat{\pi}_{BI}$ verifies that

$$M_{BI} \hat{\pi}_{BI} = \hat{\pi}_{BI}$$

and may be obtained by using the well-known power iteration method, applying it until the convergence of the iterative process

$$\begin{aligned} \hat{\pi}_k &= M_{BI} \pi_{k-1} \\ \pi_k &= \hat{\pi}_k / \max(\hat{\pi}_k), \end{aligned}$$

for $k = 1, 2, \ldots$.

In addition, it should be noted that the use of the power method for the calculation of the dominant eigenvector is especially useful when applied to sparse matrices.

3. Results

In this section, an example of a biplex network is analyzed in detail in order to highlight the possibilities offered by disposing of a centrality measure that establishes a classification of the nodes in order of importance.

For this purpose, let us consider a graph of 20 nodes, where each node represents a physical person; specifically, a player from a football team. Around this group the example is developed.

With these 20 nodes, let us proceed to construct a network with two layers that relate the nodes in a different way, taking two datasets (one for each of the layers). The calculation of the APABI centrality on this biplex network will allow us to determine the importance of each member of the team and obtain a classification of them in order of importance.

First, a layer l_1 is constructed with the 20 nodes where the relationships between the members of the team are analyzed from the point of view of social or virtual relationships. Thus, an undirected graph is constructed with an adjacency matrix A_1 in which two nodes are joined by an edge if they are related or linked through a social network. The graph of social relations between the nodes is shown in Figure 2. The data that are considered in this layer associated with each node are related to the number of messages that each person receives from their teammates in a period of time.

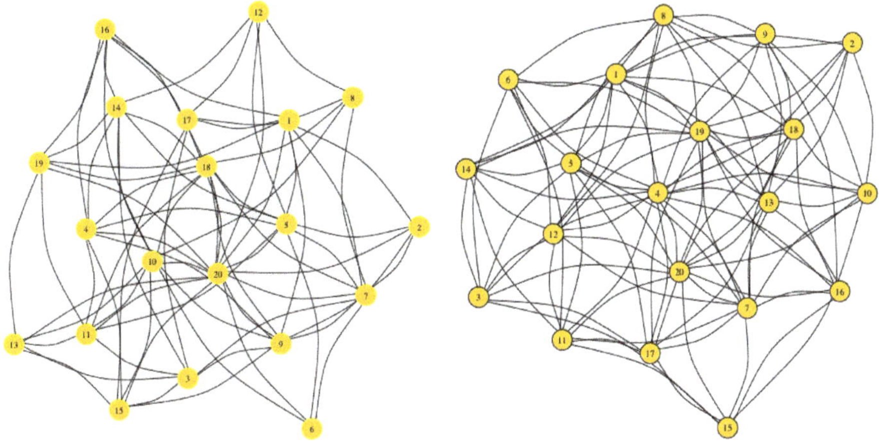

Figure 2. Graphs of the first layer (**left**) and the second one (**right**).

Secondly, a second layer l_2 is constituted from the same 20 nodes of layer l_1, but analyzing in l_2 how the players relate to each other within the game, that is, if they associate with each other in the game. Thus, two players who combine with each other or pass the ball with some assiduity during a match are connected by an edge. From these links, it is possible to build a new adjacency matrix associated with this layer, which we denote by A_2. In most team sports with a ball, each player occupies a specific position in the field of play, covering a certain area. Players (nodes) that occupy closer positions associate or relate more easily with each other than with those who are further away. For example, a defender is associated more with a midfielder than with a forward. The graph of game relations between the nodes is shown in Figure 2.

Both the links and the data associated to each of the 20 nodes of the graph are summarized in Table 1. There, the second column specifies the social links of every node while in the four column the game links between them are detailed. So, for example, the table shows that node 1 (player labeled as 1) has social interactions with the nodes $\{2, 5, 7, 9, 16, 17, 19, 20\}$ while presents strong interactions within the game with the nodes $\{2, 4, 5, 6, 9, 12, 13, 14, 18, 19\}$.

Table 1. Data associated to the biplex network constructed from the team.

Node	Social Networks Links	Messages	Game Links	Games
1	$\{2,5,7,9,16,17,19,20\}$	15	$\{2,4,5,6,9,12,13,14,18,19\}$	33
2	$\{1,5,7,9,20\}$	9	$\{1,4,8,10,13,18,19\}$	26
3	$\{7,9,11,13,14,15,17\}$	12	$\{4,5,6,12,14,15,17,20\}$	18
4	$\{5,9,11,14,15,16,18,20\}$	19	$\{1,2,3,5,6,7,8,9,10,11,12,13,14,16,17,18,20\}$	32
5	$\{1,2,4,6,7,11,12,14,18,20\}$	28	$\{1,3,4,6,7,8,11,14,17,19,20\}$	20
6	$\{5,7,10,20\}$	7	$\{1,3,4,5,8,11,12,19\}$	12
7	$\{1,2,3,5,6,8,9,18,20\}$	20	$\{4,5,10,11,12,13,15,16,17,18,19,20\}$	32
8	$\{7,10,12,17,18\}$	7	$\{2,4,5,6,9,12,13,14,18,19\}$	6
9	$\{1,2,3,4,7,10,15,18,20\}$	16	$\{1,4,8,10,13,16,18,19\}$	18
10	$\{6,8,9,11,13,14,15,16,18,20\}$	21	$\{2,4,7,9,13,15,18,19,20\}$	25
11	$\{3,4,,5,10,13,18,19,20\}$	14	$\{4,5,6,7,12,14,15,17,20\}$	24
12	$\{5,8,14,17,20\}$	8	$\{1,3,4,6,7,8,11,14,19,20\}$	18
13	$\{3,10,11,15,19,20\}$	11	$\{1,2,4,7,8,9,10,16,17,19,20\}$	6
14	$\{3,4,,5,10,12,16,18,19\}$	13	$\{1,3,4,5,8,11,12,19\}$	26
15	$\{3,4,9,10,13,17,20\}$	11	$\{3,7,10,11,16,17,20\}$	38
16	$\{1,4,10,14,17,18,19\}$	14	$\{4,7,9,13,15,18,19,20\}$	6
17	$\{1,3,8,12,15,16,20\}$	12	$\{3,4,5,7,11,13,15,18,19\}$	12
18	$\{4,5,7,8,9,10,11,14,16,19,20\}$	35	$\{1,2,4,7,8,9,10,16,17,19,20\}$	30
19	$\{1,11,13,14,16,18,20\}$	15	$\{1,2,5,6,7,8,9,10,12,13,14,16,17,18,20\}$	8
20	$\{1,2,4,5,6,7,9,10,11,12,13,15,17,18,19\}$	27	$\{3,4,5,7,10,11,12,13,15,16,18,19\}$	25

Datasets about the number of messages received during one day through social networks and the number of games that have played along the season are detailed in columns three and five, respectively. So, node 1 has received 15 messages in a day and has played 33 games in the season.

In this example, one of the advantages of working with biplex networks becomes manifest, such as the possibility of studying different relationships between the same set of nodes, analyzing the correlation between them. This example shows the advantage offered by adding what we can call a data layer in each of the multilayers of the network. We can assign the data that we consider appropriate to the specific relationships that we are representing by means of the corresponding graph. Thus, as can be seen in this example, in the layer where the social relations between the nodes are considered, we introduce the data corresponding to the number of received messages. However, in the second layer where game relations are represented, data are completely different, since now the number of games played are considered. Thus, each layer allows us to introduce one or more data sets related to the relationships of the nodes. This leads us to affirm that the inclusion of data in each of the layers enriches the nature of the problems that can be analyzed.

The objective in this example is to determine the most important or influential players within the team. For this task, two different aspects may be evaluated; on the one hand, the importance of the nodes from the point of view of the social relations that are established between them through messages, social networks, or any other virtual means. The nodes that have an intense social activity in the group create a very important union within the group, being very influential for other nodes. On the other hand, the importance of the nodes from the point of view of the game may be also evaluated, that is, which players are more important in the game, for their participation or quality. In other words, it is decisive to look for the leaders of the group, analyzing their importance from the social and technical point of view.

In order to determine the importance of the nodes of the biplex network of this example, the APABI centrality described in Section 2.3 has been calculated. Algorithm 2 has been executed using the information shown in Table 1. The numerical results shown in Table 2 are graphically displayed in Figure 3. This calculation gives us the importance of the nodes relating both layers. Figure 4 shows the final graph considering the information of two layers and the final result of the APABI centrality for each node, representing the size of each node according to its importance.

Table 2. Adapted PageBreak algorithm (APA) centrality for layers l_1 and l_2 and APA biplex (APABI) centrality for biplex network.

Node	APA Layer l_1		APA Layer l_2		APABI	
	Centrality	Ranking	Centrality	Ranking	Centrality	Ranking
1	0.05025	7	0.06394	3	0.05581	7
2	0.03110	17	0.04635	13	0.03777	16
3	0.04063	13	0.04330	14	0.04193	13
4	0.04891	9	0.08152	1	0.06517	3
5	0.07731	3	0.05134	9	0.06440	5
6	0.02494	20	0.03356	18	0.02862	20
7	0.06157	5	0.06867	2	0.06477	4
8	0.02791	19	0.03133	19	0.03071	19
9	0.05481	6	0.03965	15	0.04836	11
10	0.06530	4	0.05433	7	0.05902	6
11	0.04936	8	0.05341	8	0.05013	9
12	0.02852	18	0.04671	12	0.03727	17
13	0.03626	16	0.03407	17	0.03684	18
14	0.04776	10	0.05047	10	0.04820	12
15	0.03902	15	0.06387	4	0.05085	8
16	0.04550	12	0.02807	20	0.03781	15
17	0.04020	14	0.03830	16	0.04033	14
18	0.09182	2	0.06305	5	0.07590	2
19	0.04700	11	0.04686	11	0.04838	10
20	0.09186	1	0.06111	6	0.07775	1

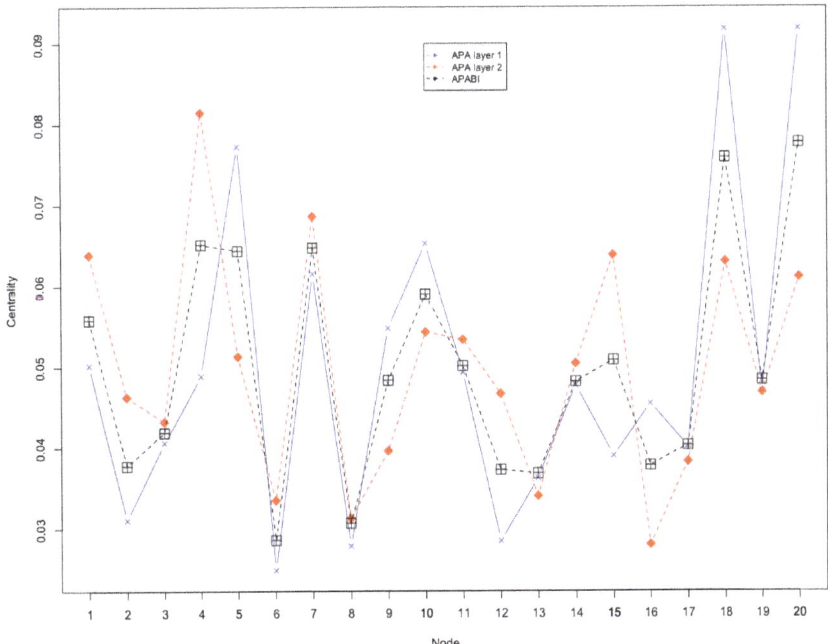

Figure 3. Centralities shown in Table 2.

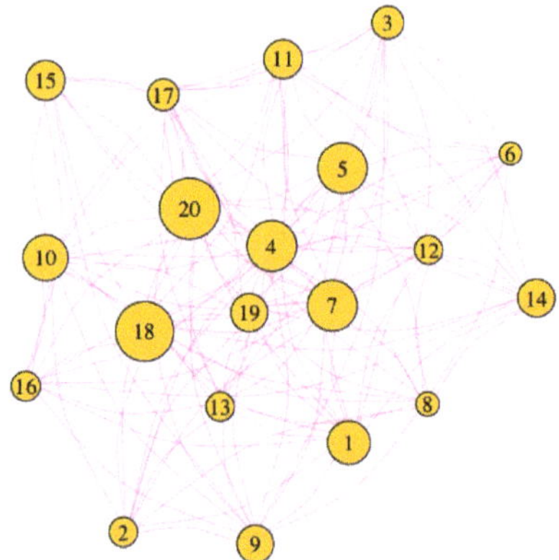

Figure 4. Adapted PageRank algorithm biplex (APABI) centrality for the example object of this study.

The APABI centrality shows that the nodes that can be classified as the most important, the leaders within the group, are nodes 20 and 18, respectively. Note that node 20, which is the most important, is not the node that receives the most messages from its colleagues, being the node 18 is the one that receives the most messages from his teammates, but is second in the ranking.

Finally, it should be noted that, as discussed in Section 2.4, the use of the power method to calculate the stationary vector of the Markov chain that forms the stochastic matrix M_{BI} provides the numerical stability needed in the implemented algorithm.

We have performed several tests with randomly generated adjacency matrices of different sizes in a range from 10 to 10,000 and we have obtained stable results. Matrices of dimension exceeding 10^5 cannot be stored in the central memory of most computers, except for sparse matrices. Consequently, the only matrix-arithmetic operation that is easily performed is a matrix-vector product. This makes possible to use this centrality algorithm for large matrices. In the scope of our research with urban network matrices, the sizes of the case studies are relatively large, with 2000–5000 nodes, approximately.

4. Discussion

In the discussion of the model presented and evaluated in the previous sections, let us pay attention to the example studied, the network of football team components. Results have been presented related to the proposed centrality measure APABI for a biplex network like this one.

We also consider that it may be relevant to calculate the nodes centrality in each of the layers, that is, separately and independently. In this way, it is possible to analyze the differences in the calculation of centrality in a single or multiplex networks, respectively, verifying if there is a certain correlation between the results obtained. We have calculated the APA centrality of the nodes in each of the layers. The numerical results obtained are shown in Table 2, where the numerical values of the centrality and the position of the nodes in the classification or ranking can be seen.

It should be noted that all the calculations have been made taking the value of the alpha parameter equal to 0.5. This means that we assign the same importance in the centrality computation to the connections of the nodes as to the data associated with them.

As it was already mentioned, the APABI centrality provides a classification of the nodes according to their importance in the biplex network studied. So, nodes 20 and 18 were the most important in the team. This can also be analyzed when the layers are considered independently.

The comparison between the APABI centrality and the APA centrality by independent layers offers us some remarkable facts. For instance, the most active nodes from the point of view of social networks are 20, 18, 10 and 5. The most important nodes from the point of view of the game are 4, 7, 1,

15. It is evident that there is no correlation between both relationships; leaders from the social point of view are not necessarily the most decisive players in the team's game. Thus, the most participative player of the team, the one who most relates to the game with his teammates, is not socially the most active nor will he be the most influential individual within the group.

We appreciate that the most important nodes in a biplex network are nodes that maintain high positions in the two rankings obtained in each of the layers. The node 20 is the one that presents a greater importance from the social point of view, however, it is not among the first three nodes that have a greater participation in the team game. On the other hand, node 18 is not as socially active as node 20, although it has a greater degree of association with its teammates in the team game than the first one and has had a greater presence in the team games, specifically participating in more than 5 games. The numerical results show that this positive rating does not compensate the high social importance of node 20, although it should be highlighted that the difference in the centrality between both nodes is very small. It can be concluded that both nodes are the team leaders.

5. Conclusions

A measure of centrality for biplex networks (APABI) based on the APA centrality for spatial networks with data has been designed and implemented following the two-layers approach for PageRank model. The advantage of having a measure of centrality of this type is twofold; on the one hand, we can determine the importance of the nodes of a network when we study various relationships between the nodes. On the other hand, we can work with several data sets associated with the nodes themselves, without any connection or relationship between them.

This measure, initially proposed for two layers of topological relationships and data, can be easily extended to any set of layers and thus add the relationships between the nodes. Its use in the study of social networks can allow us to relate various aspects of the actors, as well as their interactions. These possibilities are shown in the example studied, where the social and professional relationships of a group of people that are part of a sports group are analyzed. The importance of each node (player) is analyzed both from the social and game aspects, respectively. From the proposed centrality we determine the most influential players in the team. It is observed how the most relevant nodes in each of the layers do not have to be the most important when analyzing the related data.

Author Contributions: All authors contributed equally to this work: conceptualization, methodology, formal analysis, investigation and writing—original draft preparation.

Funding: This research is partially supported by the Spanish Government, Ministerio de Economía y Competividad, grant number TIN2017-84821-P.

Conflicts of Interest: The authors declare no conflict of interest.

Abbreviations

The following abbreviations are used in this manuscript:

APA   Adapted PageRank algorithm
APABI   Adapted PageRank algorithm biplex

References

1. Newman, M. *Networks: An Introduction*; Oxford University Press: Oxford, UK, 2010.
2. Bollobas, B. *Modern Graph Theory*; Springer: Berlin, Germany, 1998.
3. Caluset, A.; Shalizi, C.; Newman, M. Power-law distributions in empirical data. *SIAM Rev.* **2009**, *51*, 661–703. [CrossRef]
4. Porter, M. Small-world network. *Scholarpedia* **2012**, *7*, 1739. [CrossRef]
5. Boccaleti, S.; Latora, V.; Moreno, Y.; Hwang, D. Complex networks: structure and dynamics. *Phys. Rep.* **2006**, *424*, 175–308. [CrossRef]
6. Fortunato, S. Community detection in graphs. *Phys. Rep.* **2010**, *486*, 75–174. [CrossRef]

7. De Domenico, M.; Granell, C.; Porter, M.; Arenas, A. The physics of spreading processes in multilayer networks. *Nat. Phys.* **2016**, *12*, 901–906. [CrossRef]

8. De Domenico, M.; Solè-Ribalta, A.; Cozzo, E.; Kivelä, M.; Moreno, Y.; Porter, M.; Gómez, S.; Arenas, A. Mathematical formulation of multilayer networks. *Phys. Rev.* **2013**, *3*, 041022. [CrossRef]

9. Kivela, M.; Arenas, A.; Barthelemy, M.; Gleeson, J.; Moreno, Y.; Porter, M. Multilayer networks. *J. Complex Netw.* **2014**, *2*, 203–271. [CrossRef]

10. Cellai, D.; Bianconi, G. Multiplex networks with heterogeneous activities of the nodes. *Phys. Rev.* **2016**, *93*, 032302. [CrossRef]

11. De Domenico, M.; Solè-Ribalta, A.; Gómez, S.; Arenas, A. Navigability of interconnected networks under random failures. *Proc. Natl. Acad. Sci. USA* **2014**, *111*, 8351–8356. [CrossRef]

12. Padgett, J.; Ansell, C. Robust Action and the Rise of the Medici. *Am. J. Sociol.* **2016**, *98*, 1259–1319. [CrossRef]

13. Cardillo, A.; Gómez-Gardeñes, A.; Zanin, M.; Romance, M.; Papo, D.; del Pozo, F.; Boccaletti, S. Emergence of network features from multiplexity. *SIAM Rev.* **2013**, *3*, 1–122. [CrossRef] [PubMed]

14. De Domenico, M.; Lancichinetti, A.; Arenas, A.; Rosvall, M. Identifying modular flows on multilayer networks reveals highly overlapping organization in interconnected systems. *Phys. Rev. X* **2015**, *5*, 011027. [CrossRef]

15. Battiston, S.; Caldarelli, G.; May, R.; Roukny, T.; Stiglitz, J. The price of complexity in financial networks. *Proc. Natl. Acad. Sci. USA* **2016**, *113*, 10031–10036. [CrossRef] [PubMed]

16. Bentley, B.; Branicky, R.; Barnes, C.; Chew, Y.; Yemini, E.; Bullmore, E.; Vértes, P. The Multilayer Connectome of Caenorhabditis elegans. *PLOS Comput. Biol.* **2016**, *12*, e1005283. [CrossRef]

17. Sola, L.; Romance, M.; Criado, R.; Flores, J.; Garcia del Amo, A.; Boccaletti, S. Eigenvector centrality of nodes in multiplex networks. *Chaos* **2013**, *23*, 033131. [CrossRef]

18. Iacovacci, J.; Rahmede, C.; Arenas, A.; Bianconi, G. Functional Multiplex PageRank. *arXiv* **2016**, arXiv:1608.06328v2.

19. Bonacich, P. Power and centrality: A family of measures. *Am. J. Sociol.* **1987**, *92*, 1170–1182. [CrossRef]

20. Bonacich, P. Simultaneous group and individual centrality. *Soc. Netw.* **1991**, *13*, 155–168. [CrossRef]

21. Meiss, M.; Menczer, F.; Fortunato, S.; Flammini, A.; Vespignani, A. Ranking web sites with real user traffic. In Proceedings of the 2008 International Conference on Web Search and Data Mining (WSDM '08), Palo Alto, CA, USA, 11–12 February 2008; pp. 65–76.

22. Ghoshal, G.; Barabàsi, A.L. Ranking stability and super-stable nodes in complex networks. *Nat. Commun.* **2011**, *2*, 394. [CrossRef]

23. Cristelli, M.; Gabrielli, A.; Tacchella, A.; Caldarelli, G.; Pietronero, L. Measuring the intangibles: A metrics for the economic complexity of countries and products. *PLoS ONE* **2013**, *8*, e70726. [CrossRef]

24. Crucitti, P.; Latora, V.; Porta, S. Centrality measures in spatial networks of urban streets. *Phys. Rev. E* **2006**, *73*, 036125. [CrossRef] [PubMed]

25. Agryzkov, T.; Oliver, J.; Tortosa, L.; Vicent, J. An algorithm for ranking the nodes of an urban network based on the concept of PageRank vector. *Appl. Math. Comput.* **2012**, *219*, 2186–2193. [CrossRef]

26. Berkhin, P. A survey on PageRank computing. *Internet Math.* **2005**, *2*, 73–120. [CrossRef]

27. Bianconi, G. *Multilayer Networks. Structure and Functions*; Oxford University Press: Oxford, UK, 2018.

28. Halu, A.; Mondragón, R.; Panzarasa, P.; Bianconi, G. Multiplex PageRank. *PLoS ONE* **2013**, *8*, e78293. [CrossRef] [PubMed]

29. Solé-Ribalta, A.; De Domenico, M.; Gómez, S.; Arenas, A. Centrality Rankings in Multiplex Networks. In Proceedings of the 2014 ACM Conference on Web Science, Bloomington, IN, USA, 23–26 June 2014; ACM: New York, NY, USA, 2014; pp. 149–155. [CrossRef]

30. Bobadilla, J.; Ortega, F.; Hernando, A.; Gutiérrez, A. Recommender systems survey. *Knowl.-Based Syst.* **2013**, *46*, 109–132. [CrossRef]

31. Stai, E.; Kafetzoglou, S.; Tsiropoulou, E.E.; Papavassiliou, S. A Holistic Approach for Personalization, Relevance Feedback and Recommendation in Enriched Multimedia Content. *Multimedia Tools Appl.* **2018**, *77*, 283–326. [CrossRef]

32. Rabieekenari, L.; Sayrafian, K.; Baras, J. Autonomous relocation strategies for cells on wheels in environments with prohibited areas. In Proceedings of the 2017 IEEE International Conference on Communications (ICC), Paris, France, 21–25 May 2017; pp. 1–6. [CrossRef]

33. Tsiropoulou, E.; Koukas, K.; Papavassiliou, S. A Socio-physical and Mobility-Aware Coalition Formation Mechanism in Public Safety Networks. *EAI Endorsed Trans. Future Internet* **2018**, *4*. [CrossRef]
34. Pedroche, F.; Romance, M.; Criado, R. A biplex approach to PageRank centrality: From classic to multiplex networks. *Chaos* **2016**, *26*, 065301. [CrossRef]
35. Agryzkov, T.; Tortosa, L.; Vicent, J.; Wilson, R. A centrality measure for urban networks based on the eigenvector centrality concept. *Environ. Plan. B* **2017**, *291*, 14–29. [CrossRef]
36. Page, L.; Brin, S.; Motwani, R.; Winogrand, T. *The Pagerank Citation Ranking: Bringing Order to the Web*; Technical Report 1999-66; Stanford InfoLab: Stanford, CA, USA 1999; Volume 66.
37. Pedroche, F. Métodos de cálculo del vector PageRank. *Bol. Soc. Esp. Mat. Apl.* **2007**, *39*, 7–30.
38. Agryzkov, T.; Pedroche, F.; Tortosa, L.; Vicent, J. Combining the Two-Layers PageRank Approach with the APA Centrality in Networks with Data. *Int. J. Geo-Inform.* **2018**, *7*. [CrossRef]
39. Datta, B. *Numerical Linear Algebra and Applications*; Brooks/Cole Publishing Company: Pacific Grove, CA, USA, 1995.

Article

Mei Symmetry and Invariants of Quasi-Fractional Dynamical Systems with Non-Standard Lagrangians

Yi Zhang [1,*] and Xue-Ping Wang [2]

[1]   College of Civil Engineering, Suzhou University of Science and Technology, Suzhou 215011, China
[2]   College of Mathematics and Physics, Suzhou University of Science and Technology, Suzhou 215009, China
*   Correspondence: zhy@mail.usts.edu.cn

Received: 15 July 2019; Accepted: 15 August 2019; Published: 18 August 2019

Abstract: Non-standard Lagrangians play an important role in the systems of non-conservative dynamics or nonlinear differential equations, quantum field theories, etc. This paper deals with quasi-fractional dynamical systems from exponential non-standard Lagrangians and power-law non-standard Lagrangians. Firstly, the definition, criterion, and corresponding new conserved quantity of Mei symmetry in this system are presented and studied. Secondly, considering that a small disturbance is applied on the system, the differential equations of the disturbed motion are established, the definition of Mei symmetry and corresponding criterion are given, and the new adiabatic invariants led by Mei symmetry are proposed and proved. Examples also show the validity of the results.

Keywords: Mei symmetry; conserved quantity; adiabatic invariant; quasi-fractional dynamical system; non-standard Lagrangians

---

1. Introduction

The study of symmetry and invariants for non-conservative or nonlinear dynamics is of great significance. It is also a frontier research field of analytical mechanics. In a classical sense, the symmetries we refer to mainly include Noether symmetries [1] and Lie symmetries [2,3]. Noether symmetry and Lie symmetry are two different symmetries. After infinitesimal transformation, the former means the invariant property of the Hamilton action functional, and the latter means the invariant property of the differential equation. Unlike Noether symmetry or Lie symmetry, Mei proposed a new symmetry called form invariance in 2000 [4]. Form invariance, also known as Mei symmetry, refers to an invariant property, that is, the dynamical functions (such as Lagrangian, Hamiltonian, Birkhoffian, generalized force, etc.) that appear in the dynamical equations of the mechanical system still satisfy the original equations after the infinitesimal transformation. Under certain conditions, symmetry can lead to invariants, which are also called conserved quantities. Noether symmetry, Lie symmetry, and Mei symmetry of dynamical systems described by standard Lagrangian may lead to Noether conserved quantities or Mei conserved quantities [5], etc. Conserved quantities can also be independent of Lagrangian. For example, conserved quantities can be directly constructed from Lie symmetry neither utilizing Lagrangian nor Hamiltonian, or can be formulated for systems of differential equations by using symmetries and adjoint symmetries together regardless of the existence of a Lagrangian, see [5–8] and references therein. So far, much progress has been made in the study of the symmetries and corresponding invariants [9–22]. However, there are few reports on the symmetries and invariants of dynamical systems based on non-standard Lagrangians.

The concept of non-standard Lagrangian was first mentioned in Arnold's works in 1978 as non-natural Lagrangian [23], but it was ignored due to the lack of Hamilton form corresponding to it. Until 1984, when discussing the region adaptability of classical theories in Yang–Mills quantum field theory [24], it was found that non-standard Lagrangians were directly related to the color constraint problem, which led to their renewed attention. The advantage of non-standard Lagrangians is that it can better describe nonlinear problems and plays an important role in non-conservative systems, dissipative systems, quantum field theory, etc. [25–34].

Fractional calculus can better describe natural phenomena and engineering problems [35–37]. Since Riewe [38,39] introduced fractional calculus into the modeling of non-conservative systems, fractional Lagrangian mechanics, fractional Hamiltonian mechanics and fractional Birkhoffian mechanics have been proposed and studied, and important progress has been made in fractional dynamics modeling, analysis, and calculation, see for example [40–47] and references therein. In 2005, El-Nabulsi proposed the fractional action-like variational approach to study non-conservative dynamical problem, in which the action is constructed by using the Riemann-Liouville definition of fractional integral [48,49], and extended it to the case of non-standard Lagrangians [28,50]. Considering the characteristic of fractional action-like variational approach, we call the non-conservative model obtained in this way as quasi-fractional order dynamical system. Here, we propose and study Mei symmetry and its invariants for the quasi-fractional order dynamical system with non-standard Lagrangians. New conserved quantities and new adiabatic invariants are derived from Mei symmetry of the quasi-fractional dynamical systems.

2. Mei Symmetry and Invariants of Quasi-Fractional Dynamical System Based on Exponential Lagrangians

For the quasi-fractional dynamical system whose action functional depends on exponential Lagrangian, the Euler–Lagrange equations that are derived in Appendix A can be expressed as

$$(t-\tau)^{\alpha-1}\exp L\left(\frac{\partial L}{\partial q_s} - \frac{\mathrm{d}}{\mathrm{d}\tau}\frac{\partial L}{\partial \dot{q}_s} - \frac{\partial L}{\partial \dot{q}_s}\frac{\mathrm{d}L}{\mathrm{d}\tau} + \frac{\alpha-1}{t-\tau}\frac{\partial L}{\partial \dot{q}_s}\right) = 0, \ (s = 1,2,\cdots,n), \tag{1}$$

where $q_s(s = 1,2,\cdots,n)$ are the generalized coordinates, $L = L\left(\tau, q_s, \dot{q}_s\right)$ is the standard Lagrangian, $0 < \alpha \leq 1$, τ is the intrinsic time, t is the observer time, and τ is not equal to t.

Let us introduce the infinitesimal transformations as

$$\tau^* = \tau + \varepsilon\varsigma_0\left(\tau, q_k, \dot{q}_k\right), \ q_s^*(\tau^*) = q_s(\tau) + \varepsilon\xi_s\left(\tau, q_k, \dot{q}_k\right), \ (s = 1,2,\cdots,n; k = 1,2,\cdots,n), \tag{2}$$

where ε is a small parameter, ς_0 and ξ_s are the infinitesimals. After the transformation of Equation (2), $\exp L$ is transformed into the following form

$$\exp L^* = \exp L\left(\tau^*, q_s^*, \frac{\mathrm{d}q_s^*}{\mathrm{d}\tau^*}\right) = \exp L\left(\tau, q_s, \dot{q}_s\right) + \varepsilon X^{(1)}(\exp L) + O\left(\varepsilon^2\right), \tag{3}$$

where $X^{(1)}$ is the first extension of the infinitesimal generator X, that is [4]

$$X = \varsigma_0\frac{\partial}{\partial\tau} + \xi_s\frac{\partial}{\partial q_s}, \ X^{(1)} = \varsigma_0\frac{\partial}{\partial\tau} + \xi_s\frac{\partial}{\partial q_s} + \left(\dot{\xi}_s - \dot{q}_s\dot{\varsigma}_0\right)\frac{\partial}{\partial\dot{q}_s}. \tag{4}$$

If L is replaced with L^*, Equation (1) still holds, namely

$$(t-\tau)^{\alpha-1}\exp L^*\left(\frac{\partial L^*}{\partial q_s} - \frac{\mathrm{d}}{\mathrm{d}\tau}\frac{\partial L^*}{\partial \dot{q}_s} - \frac{\partial L^*}{\partial \dot{q}_s}\frac{\mathrm{d}L^*}{\mathrm{d}\tau} + \frac{\alpha-1}{t-\tau}\frac{\partial L^*}{\partial \dot{q}_s}\right) = 0, \ (s = 1,2,\cdots,n), \tag{5}$$

then this invariance is called Mei symmetry of quasi-fractional dynamical system (1). Substituting the formula (3) into Equation (5), and considering Equation (1), we have

$$(t-\tau)^{\alpha-1}\exp L\left(\frac{\partial X^{(1)}(L)}{\partial q_s} - \frac{\mathrm{d}}{\mathrm{d}\tau}\frac{\partial X^{(1)}(L)}{\partial \dot{q}_s} - \frac{\partial X^{(1)}(L)}{\partial \dot{q}_s}\frac{\mathrm{d}L}{\mathrm{d}\tau}\right.$$
$$\left. -\frac{\partial L}{\partial \dot{q}_s}\frac{\mathrm{d}X^{(1)}(L)}{\mathrm{d}\tau} + \frac{\alpha-1}{t-\tau}\frac{\partial X^{(1)}(L)}{\partial \dot{q}_s}\right) = 0, \quad (s = 1,2,\cdots,n). \tag{6}$$

Equation (6) is the criterion for Mei symmetry of system (1).

Theorem 1. *For the quasi-fractional dynamical system (1), if there is a gauge function $G = G\left(\tau, q_k, \dot{q}_k\right)$ such that the structural equation*

$$\left(\frac{1-\alpha}{t-\tau}\varsigma_0 + \dot{\varsigma}_0\right)X^{(1)}(\exp L)(t-\tau)^{\alpha-1} + X^{(1)}\left[X^{(1)}(\exp L)\right](t-\tau)^{\alpha-1} + \dot{G} = 0 \tag{7}$$

holds, the Mei symmetry directly leads to the new conserved quantity

$$I_0 = (t-\tau)^{\alpha-1}X^{(1)}(\exp L)\varsigma_0 + (t-\tau)^{\alpha-1}\frac{\partial X^{(1)}(\exp L)}{\partial \dot{q}_s}\left(\xi_s - \dot{q}_s\varsigma_0\right) + G = \mathrm{const.} \tag{8}$$

Proof.

$$\frac{\mathrm{d}I_0}{\mathrm{d}\tau} = \frac{1-\alpha}{t-\tau}(t-\tau)^{\alpha-1}X^{(1)}(\exp L)\varsigma_0 + (t-\tau)^{\alpha-1}\frac{\mathrm{d}X^{(1)}(\exp L)}{\mathrm{d}\tau}\varsigma_0 + (t-\tau)^{\alpha-1}X^{(1)}(\exp L)\dot{\varsigma}_0$$
$$+\frac{1-\alpha}{t-\tau}(t-\tau)^{\alpha-1}\frac{\partial X^{(1)}(\exp L)}{\partial \dot{q}_s}\left(\xi_s - \dot{q}_s\varsigma_0\right) + (t-\tau)^{\alpha-1}\frac{\mathrm{d}}{\mathrm{d}\tau}\frac{\partial X^{(1)}(\exp L)}{\partial \dot{q}_s}\left(\xi_s - \dot{q}_s\varsigma_0\right)$$
$$+(t-\tau)^{\alpha-1}\frac{\partial X^{(1)}(\exp L)}{\partial \dot{q}_s}\left(\dot{\xi}_s - \dot{q}_s\dot{\varsigma}_0 - \ddot{q}_s\varsigma_0\right) - (t-\tau)^{\alpha-1}X^{(1)}(\exp L)\left(\frac{1-\alpha}{t-\tau}\varsigma_0 + \dot{\varsigma}_0\right)$$
$$-(t-\tau)^{\alpha-1}X^{(1)}\left[X^{(1)}(\exp L)\right]$$
$$=\left[-\frac{\partial X^{(1)}(L)}{\partial q_s} + \frac{\partial X^{(1)}(L)}{\partial \dot{q}_s}\frac{\mathrm{d}L}{\mathrm{d}\tau} + \frac{\partial L}{\partial \dot{q}_s}\frac{\mathrm{d}X^{(1)}(L)}{\mathrm{d}\tau} + \frac{\mathrm{d}}{\mathrm{d}\tau}\frac{\partial X^{(1)}(L)}{\partial \dot{q}_s} + \frac{1-\alpha}{t-\tau}\frac{\partial X^{(1)}(L)}{\partial \dot{q}_s}\right]\times$$
$$\times\left(\xi_s - \dot{q}_s\varsigma_0\right)(t-\tau)^{\alpha-1}\exp L + \left(\frac{1-\alpha}{t-\tau}\varsigma_0 + \dot{\varsigma}_0\right)X^{(1)}(\exp L)(t-\tau)^{\alpha-1}$$
$$+X^{(1)}\left[X^{(1)}(\exp L)\right](t-\tau)^{\alpha-1} + \left(-\frac{\partial L}{\partial q_s} + \frac{\mathrm{d}}{\mathrm{d}\tau}\frac{\partial L}{\partial \dot{q}_s} + \frac{\partial L}{\partial \dot{q}_s}\frac{\mathrm{d}L}{\mathrm{d}\tau} + \frac{1-\alpha}{t-\tau}\frac{\partial L}{\partial \dot{q}_s}\right)\times$$
$$\times\left(\xi_s - \dot{q}_s\varsigma_0\right)(t-\tau)^{\alpha-1}X^{(1)}(\exp L) + \dot{G}. \tag{9}$$

Substituting Equations (1) and (6) into the formula (9), and using Equation (7), we obtain

$$\frac{\mathrm{d}I_0}{\mathrm{d}\tau} = \left\{\left(\frac{1-\alpha}{t-\tau}\varsigma_0 + \dot{\varsigma}_0\right)X^{(1)}(\exp L) + X^{(1)}\left[X^{(1)}(\exp L)\right]\right\}(t-\tau)^{\alpha-1} + \dot{G} = 0. \tag{10}$$

Thus, we get the desired result. $\square$

The new conserved quantity (8) is called Mei conserved quantity. Since the system is not disturbed, it is an exact invariant. However, in nature and engineering, it is often affected by disturbing forces. If the system is affected by small disturbance $v Q_s$, its Mei symmetry and the corresponding conserved quantity (8) will change correspondingly. The infinitesimals of transformations (2) without disturbance is denoted as ς_0^0, ξ_s^0, while the infinitesimals are changed into ς_0, ξ_s when disturbed, and we have

$$\varsigma_0 = \varsigma_0^0 + v\varsigma_0^1 + v^2\varsigma_0^2 + \cdots, \quad \xi_s = \xi_s^0 + v\xi_s^1 + v^2\xi_s^2 + \cdots, \quad (s = 1,2,\cdots,n). \tag{11}$$

Meanwhile, we let G^0 represent the gauge function without disturbance, and G represent the gauge function of the disturbed system, which is the small perturbation on the basis of G^0, i.e.,

$$G = G^0 + v G^1 + v^2 G^2 + \cdots. \tag{12}$$

If a small disturbance vQ_s is applied, Equation (1) is changed to

$$(t-\tau)^{\alpha-1}\exp L\left(\frac{\partial L}{\partial q_s}-\frac{\mathrm{d}}{\mathrm{d}\tau}\frac{\partial L}{\partial \dot{q}_s}-\frac{\partial L}{\partial \dot{q}_s}\frac{\mathrm{d}L}{\mathrm{d}\tau}+\frac{\alpha-1}{t-\tau}\frac{\partial L}{\partial \dot{q}_s}\right)=vQ_s. \tag{13}$$

Accordingly, Equation (6) is changed to

$$(t-\tau)^{\alpha-1}\exp L\left[\frac{\partial X^{(1)}(L)}{\partial q_s}-\frac{\mathrm{d}}{\mathrm{d}\tau}\frac{\partial X^{(1)}(L)}{\partial \dot{q}_s}-\frac{\partial X^{(1)}(L)}{\partial \dot{q}_s}\frac{\mathrm{d}L}{\mathrm{d}\tau}\right.$$
$$\left.-\frac{\partial L}{\partial \dot{q}_s}\frac{\mathrm{d}X^{(1)}(L)}{\mathrm{d}\tau}+\frac{\alpha-1}{t-\tau}\frac{\partial X^{(1)}(L)}{\partial \dot{q}_s}\right]=vX^{(1)}(Q_s),\ (s=1,2,\cdots,n). \tag{14}$$

Substituting the formulae (11) into Equation (14), we get

$$(t-\tau)^{\alpha-1}v^m\exp L\left[\frac{\partial X_m^{(1)}(L)}{\partial q_s}-\frac{\mathrm{d}}{\mathrm{d}\tau}\frac{\partial X_m^{(1)}(L)}{\partial \dot{q}_s}-\frac{\partial X_m^{(1)}(L)}{\partial \dot{q}_s}\frac{\mathrm{d}L}{\mathrm{d}\tau}\right.$$
$$\left.-\frac{\partial L}{\partial \dot{q}_s}\frac{\mathrm{d}X_m^{(1)}(L)}{\mathrm{d}\tau}+\frac{\alpha-1}{t-\tau}\frac{\partial X_m^{(1)}(L)}{\partial \dot{q}_s}\right]=v^{m+1}X_m^{(1)}(Q_s),\ (s=1,2,\cdots,n). \tag{15}$$

where

$$X^{(1)}=v^mX_m^{(1)},\ X_m^{(1)}=\varsigma_0^m\frac{\partial}{\partial \tau}+\xi_s^m\frac{\partial}{\partial q_s}+\left(\dot{\xi}_s^m-\dot{q}_s\dot{\varsigma}_0^m\right)\frac{\partial}{\partial \dot{q}_s}. \tag{16}$$

As a result, we have

Theorem 2. *If the quasi-fractional dynamical system (1) is disturbed by a small disturbance vQ_s, and there is a gauge function $G=G\left(\tau,q_k,\dot{q}_k\right)$ such that the structural equation*

$$\left(\frac{1-\alpha}{t-\tau}\varsigma_0^m+\dot{\varsigma}_0^m\right)X_m^{(1)}(\exp L)(t-\tau)^{\alpha-1}+X_m^{(1)}\left[X_m^{(1)}(\exp L)\right](t-\tau)^{\alpha-1}+\dot{G}^m$$
$$-\left[X_{m-1}^{(1)}(Q_s)+Q_sX_{m-1}^{(1)}(L)\right]\left(\xi_s^{m-1}-\dot{q}_s\varsigma_0^{m-1}\right)=0,\ (s=1,2,\cdots,n;m=0,1,2,\cdots), \tag{17}$$

holds, where $G=\sum\limits_{m=0}^{z}v^mG^m$ and $\varsigma_0^{-1}=\xi_s^{-1}=0$, the Mei symmetry directly leads to the new adiabatic invariant

$$I_z=\sum_{m=0}^{z}v^m\left[(t-\tau)^{\alpha-1}X_m^{(1)}(\exp L)\varsigma_0^m+(t-\tau)^{\alpha-1}\frac{\partial X_m^{(1)}(\exp L)}{\partial \dot{q}_s}\left(\xi_s^m-\dot{q}_s\varsigma_0^m\right)+G^m\right]. \tag{18}$$

Proof. By using Equations (13), (15), and (17), we have

$$
\begin{aligned}
\frac{dI_z}{d\tau} &= \sum_{m=0}^{z} v^m \Big\{ \tfrac{1-\alpha}{t-\tau}(t-\tau)^{\alpha-1} X_m^{(1)}(\exp L)\varsigma_0^m + (t-\tau)^{\alpha-1}\frac{dX_m^{(1)}(\exp L)}{d\tau}\varsigma_0^m \\
&\quad +(t-\tau)^{\alpha-1} X_m^{(1)}(\exp L)\dot\varsigma_0^m + \tfrac{1-\alpha}{t-\tau}(t-\tau)^{\alpha-1}\frac{\partial X_m^{(1)}(\exp L)}{\partial \dot q_s}\big(\xi_s^m - \dot q_s\varsigma_0^m\big) \\
&\quad +(t-\tau)^{\alpha-1}\frac{d}{d\tau}\frac{\partial X_m^{(1)}(\exp L)}{\partial \dot q_s}\big(\xi_s^m - \dot q_s\varsigma_0^m\big) + (t-\tau)^{\alpha-1}\frac{\partial X_m^{(1)}(\exp L)}{\partial \dot q_s}\big(\dot\xi_s^m - \dot q_s\dot\varsigma_0^m - \ddot q_s\varsigma_0^m\big) + \dot G^m \Big\} \\
&= \sum_{m=0}^{z} v^m \Big\{ \Big[-\frac{\partial X_m^{(1)}(L)}{\partial q_s} + \frac{\partial X_m^{(1)}(L)}{\partial \dot q_s}\frac{dL}{d\tau} + \frac{\partial L}{\partial \dot q_s}\frac{dX_m^{(1)}(L)}{d\tau} + \frac{d}{d\tau}\frac{\partial X_m^{(1)}(L)}{\partial \dot q_s} + \frac{1-\alpha}{t-\tau}\frac{\partial X_m^{(1)}(L)}{\partial \dot q_s} \Big] \times \\
&\quad \times\big(\xi_s^m - \dot q_s\varsigma_0^m\big)(t-\tau)^{\alpha-1}\exp L + \Big(\tfrac{1-\alpha}{t-\tau}\varsigma_0^m + \dot\varsigma_0^m\Big) X_m^{(1)}(\exp L)(t-\tau)^{\alpha-1} \\
&\quad + X_m^{(1)}\Big[X_m^{(1)}(\exp L)\Big](t-\tau)^{\alpha-1} - \Big(\frac{\partial L}{\partial q_s} - \frac{d}{d\tau}\frac{\partial L}{\partial \dot q_s} - \frac{\partial L}{\partial \dot q_s}\frac{dL}{d\tau} - \frac{1-\alpha}{t-\tau}\frac{\partial L}{\partial \dot q_s}\Big)\times \\
&\quad \times\big(\xi_s^m - \dot q_s\varsigma_0^m\big)(t-\tau)^{\alpha-1} X_m^{(1)}(L)\exp L + \dot G^m \Big\} \\
&= \sum_{m=0}^{z} v^m \Big\{ -v X_m^{(1)}(Q_s)\big(\xi_s^m - \dot q_s\varsigma_0^m\big) + X_{m-1}^{(1)}(Q_s)\big(\xi_s^{m-1} - \dot q_s\varsigma_0^{m-1}\big) \\
&\quad + Q_s X_{m-1}^{(1)}(L)\big(\xi_s^{m-1} - \dot q_s\varsigma_0^{m-1}\big) - v Q_s X_m^{(1)}(L)\big(\xi_s^m - \dot q_s\varsigma_0^m\big) \Big\} \\
&= -v^{z+1}\Big[X_z^{(1)}(Q_s) + Q_s X_z^{(1)}(L) \Big]\big(\xi_s^z - \dot q_s\varsigma_0^z\big).
\end{aligned}
\tag{19}
$$

According to the definition of adiabatic invariant [51], I_z is an adiabatic invariant of order z. This completes the proof. □

Example 1. *Considering the nonlinear dynamical system, its action functional based on exponential Lagrangian is*

$$
S = \frac{1}{\Gamma(\alpha)}\int_{t_1}^{t_2} \exp\big[L(\tau, q_s, \dot q_s)\big](t-\tau)^{\alpha-1}d\tau,
\tag{20}
$$

where $L = \tau q\dot q$.

Equation (1) gives

$$
(t-\tau)^{\alpha-1}\exp(\tau q\dot q)\Big[\tau q\Big(\frac{\alpha-1}{t-\tau} - q\dot q - \tau\dot q^2 - \tau q\ddot q\Big) - q\Big] = 0.
\tag{21}
$$

By calculation, we have

$$
X_0^{(1)}(L) = q\dot q\varsigma_0^0 + \tau\dot q\xi^0 + \tau q\Big(\dot\xi^0 - \dot q\dot\varsigma_0^0\Big),
\tag{22}
$$

$$
X_0^{(1)}(\exp L) = \exp(\tau q\dot q)\Big[q\dot q\varsigma_0^0 + \tau\dot q\xi^0 + \tau q\Big(\dot\xi^0 - \dot q\dot\varsigma_0^0\Big)\Big].
\tag{23}
$$

If we take

$$
\varsigma_0^0 = \tau, \ \xi^0 = \frac{1}{q},
\tag{24}
$$

then we have

$$
X_0^{(1)}(L) = 0, \ X_0^{(1)}(\exp L) = 0.
\tag{25}
$$

According to the criterion (6), the infinitesimals (24) correspond to Mei symmetry. Substituting (24) into Equation (7), we get

$$
G^0 = \tau\exp(\tau q\dot q)(t-\tau)^{\alpha-1}.
\tag{26}
$$

From Theorem 1, we have

$$
I_0 = \tau\exp(\tau q\dot q)(t-\tau)^{\alpha-1} = \text{const.}
\tag{27}
$$

Let the small disturbance be

$$vQ = v q\dot{q}\exp\left(q^2/2\right).\tag{28}$$

The differential equation of the disturbed motion is

$$(t-\tau)^{\alpha-1}\exp\left(\tau q\dot{q}\right)\left[\tau q\left(\frac{\alpha-1}{t-\tau}-q\dot{q}-\tau\dot{q}^2-\tau q\ddot{q}\right)-q\right]=v q\dot{q}\exp\left(q^2/2\right).\tag{29}$$

Take

$$\varsigma_0^1 = \tau, \ \xi^1 = \frac{1}{q},\tag{30}$$

then we have

$$X_1^{(1)}(L)=0,\ X_1^{(1)}(\exp L)=0,\ X_0^{(1)}(Q)=X_1^{(1)}(Q)=0.\tag{31}$$

According to the criterion (15), the infinitesimals (30) correspond to Mei symmetry. Substituting (30) into Equation (17), we have

$$G^1 = \tau\exp\left(\tau q\dot{q}\right)(t-\tau)^{\alpha-1}+v\int\exp\left(q^2/2\right)\mathrm{d}q.\tag{32}$$

By Theorem 2, we obtain

$$I_1 = \tau\exp\left(\tau q\dot{q}\right)(t-\tau)^{\alpha-1}+v\left[\tau\exp\left(\tau q\dot{q}\right)(t-\tau)^{\alpha-1}+v\int\exp\left(q^2/2\right)\mathrm{d}q\right].\tag{33}$$

The formula (33) is an adiabatic invariant led by Mei symmetry.

3. Mei Symmetry and Invariants of Quasi-Fractional Dynamical System Based on Power-Law Lagrangians

For the quasi-fractional dynamical system whose action functional depends on power-law Lagrangian, the Euler–Lagrange equations that are derived in Appendix B can be expressed as

$$(1+\gamma)(t-\tau)^{\alpha-1}L^\gamma\left(\frac{\partial L}{\partial q_s}-\frac{\mathrm{d}}{\mathrm{d}\tau}\frac{\partial L}{\partial\dot{q}_s}-\frac{\gamma}{L}\frac{\partial L}{\partial\dot{q}_s}\frac{\mathrm{d}L}{\mathrm{d}\tau}+\frac{\alpha-1}{t-\tau}\frac{\partial L}{\partial\dot{q}_s}\right)=0,\ (s=1,2,\cdots,n),\tag{34}$$

where γ is not equal to -1.

After the transformation of (2), $L^{1+\gamma}$ is transformed into the following form

$$L^{*1+\gamma}=L^{1+\gamma}\left(\tau^*,q_s^*,\frac{\mathrm{d}q_s^*}{\mathrm{d}\tau^*}\right)=L^{1+\gamma}\left(\tau,q_s,\dot{q}_s\right)+\varepsilon X^{(1)}\left(L^{1+\gamma}\right)+O\left(\varepsilon^2\right).\tag{35}$$

If L is replaced with L^*, Equation (34) still holds, namely

$$(1+\gamma)(t-\tau)^{\alpha-1}L^{*\gamma}\left(\frac{\partial L^*}{\partial q_s}-\frac{\mathrm{d}}{\mathrm{d}\tau}\frac{\partial L^*}{\partial\dot{q}_s}-\frac{\gamma}{L^*}\frac{\partial L^*}{\partial\dot{q}_s}\frac{\mathrm{d}L^*}{\mathrm{d}\tau}+\frac{\alpha-1}{t-\tau}\frac{\partial L^*}{\partial\dot{q}_s}\right)=0,\ (s=1,2,\cdots,n).\tag{36}$$

then this invariance is called Mei symmetry of quasi-fractional dynamical system (34). Substituting the formula (35) into Equation (36), and considering Equation (34), we have

$$\begin{aligned}(1+\gamma)(t-\tau)^{\alpha-1}L^\gamma\Bigg[&\frac{\partial X^{(1)}(L)}{\partial q_s}-\frac{\mathrm{d}}{\mathrm{d}\tau}\frac{\partial X^{(1)}(L)}{\partial\dot{q}_s}-\frac{\gamma}{L}\frac{\mathrm{d}L}{\mathrm{d}\tau}\frac{\partial X^{(1)}(L)}{\partial\dot{q}_s}-\frac{\gamma}{L}\frac{\partial L}{\partial\dot{q}_s}\frac{\mathrm{d}X^{(1)}(L)}{\mathrm{d}\tau}\\&+\frac{\gamma}{L^2}\frac{\partial L}{\partial\dot{q}_s}\frac{\mathrm{d}L}{\mathrm{d}\tau}X^{(1)}(L)+\frac{\alpha-1}{t-\tau}\frac{\partial X^{(1)}(L)}{\partial\dot{q}_s}\Bigg]=0,\ (s=1,2,\cdots,n).\end{aligned}\tag{37}$$

Equation (37) is the criterion for Mei symmetry of system (34). Hence, we have

Theorem 3. *For the quasi-fractional dynamical system (34), if there is a gauge function* $G = G\left(\tau, q_k, \dot{q}_k\right)$ *such that the structural equation*

$$\left(\frac{1-\alpha}{t-\tau}\varsigma_0 + \dot{\varsigma}_0\right)X^{(1)}\left(L^{1+\gamma}\right) + X^{(1)}\left\{X^{(1)}\left(L^{1+\gamma}\right)\right\} + (t-\tau)^{1-\alpha}\dot{G} = 0 \tag{38}$$

holds, the Mei symmetry directly leads to the new conserved quantity

$$I_0 = (t-\tau)^{\alpha-1}X^{(1)}\left(L^{1+\gamma}\right)\varsigma_0 + (t-\tau)^{\alpha-1}\frac{\partial X^{(1)}\left(L^{1+\gamma}\right)}{\partial \dot{q}_s}\left(\xi_s - \dot{q}_s\varsigma_0\right) + G = \text{const.} \tag{39}$$

Proof.

$$\begin{aligned}
\frac{\mathrm{d}I_0}{\mathrm{d}t} &= \frac{1-\alpha}{t-\tau}(t-\tau)^{\alpha-1}X^{(1)}\left(L^{1+\gamma}\right)\varsigma_0 + (t-\tau)^{\alpha-1}\frac{\mathrm{d}X^{(1)}\left(L^{1+\gamma}\right)}{\mathrm{d}\tau}\varsigma_0 + (t-\tau)^{\alpha-1}X^{(1)}\left(L^{1+\gamma}\right)\dot{\varsigma}_0 \\
&\quad + \frac{1-\alpha}{t-\tau}(t-\tau)^{\alpha-1}\frac{\partial X^{(1)}\left(L^{1+\gamma}\right)}{\partial \dot{q}_s}\left(\xi_s - \dot{q}_s\varsigma_0\right) + (t-\tau)^{\alpha-1}\frac{\mathrm{d}}{\mathrm{d}\tau}\frac{\partial X^{(1)}\left(L^{1+\gamma}\right)}{\partial \dot{q}_s}\left(\xi_s - \dot{q}_s\varsigma_0\right) \\
&\quad + (t-\tau)^{\alpha-1}\frac{\partial X^{(1)}\left(L^{1+\gamma}\right)}{\partial \dot{q}_s}\left(\dot{\xi}_s - \dot{q}_s\dot{\varsigma}_0 - \ddot{q}_s\varsigma_0\right) + \dot{G} \\
&= \left[-\frac{\partial X^{(1)}(L)}{\partial q_s} + \frac{\mathrm{d}}{\mathrm{d}\tau}\frac{\partial X^{(1)}(L)}{\partial \dot{q}_s} + \frac{\gamma}{L}\frac{\mathrm{d}L}{\mathrm{d}\tau}\frac{\partial X^{(1)}(L)}{\partial \dot{q}_s} + \frac{\gamma}{L}\frac{\partial L}{\partial \dot{q}_s}\frac{\mathrm{d}X^{(1)}(L)}{\mathrm{d}\tau} - \frac{\gamma}{L^2}\frac{\partial L}{\partial \dot{q}_s}\frac{\mathrm{d}L}{\mathrm{d}\tau}X^{(1)}(L)\right. \\
&\quad \left. + \frac{1-\alpha}{t-\tau}\frac{\partial X^{(1)}(L)}{\partial \dot{q}_s}\right](1+\gamma)(t-\tau)^{\alpha-1}L^{\gamma}\left(\xi_s - \dot{q}_s\varsigma_0\right) + \left(\frac{1-\alpha}{t-\tau}\varsigma_0 + \dot{\varsigma}_0\right)X^{(1)}\left(L^{1+\gamma}\right)(t-\tau)^{\alpha-1} \\
&\quad + \left(-\frac{\partial L}{\partial q_s} + \frac{\mathrm{d}}{\mathrm{d}\tau}\frac{\partial L}{\partial \dot{q}_s} + \frac{\gamma}{L}\frac{\partial L}{\partial \dot{q}_s}\frac{\mathrm{d}L}{\mathrm{d}\tau} + \frac{1-\alpha}{t-\tau}\frac{\partial L}{\partial \dot{q}_s}\right)(1+\gamma)(t-\tau)^{\alpha-1}X^{(1)}(L^{\gamma})\left(\xi_s - \dot{q}_s\varsigma_0\right) \\
&\quad + X^{(1)}(L)X^{(1)}(L)(1+\gamma)\gamma(t-\tau)^{\alpha-1}L^{\gamma-1} + X^{(1)}\left[X^{(1)}(L)\right](1+\gamma)(t-\tau)^{\alpha-1}L^{\gamma} + \dot{G}.
\end{aligned} \tag{40}$$

Substituting Equations (34) and (37) into the formula (40), and using Equation (38), we obtain

$$\frac{\mathrm{d}I_0}{\mathrm{d}t} = \left(\frac{1-\alpha}{t-\tau}\varsigma_0 + \dot{\varsigma}_0\right)X^{(1)}\left(L^{1+\gamma}\right)(t-\tau)^{\alpha-1} + X^{(1)}\left[X^{(1)}\left(L^{1+\gamma}\right)\right](t-\tau)^{\alpha-1} + \dot{G} = 0. \tag{41}$$

Thus, we get the desired result. □

Mei conserved quantity (39) is an exact invariant for the quasi-fractional dynamical systems (34). If a small disturbance vQ_s is applied, Equation (34) is changed to

$$(1+\gamma)(t-\tau)^{\alpha-1}L^{\gamma}\left(\frac{\partial L}{\partial q_s} - \frac{\mathrm{d}}{\mathrm{d}\tau}\frac{\partial L}{\partial \dot{q}_s} - \frac{\gamma}{L}\frac{\partial L}{\partial \dot{q}_s}\frac{\mathrm{d}L}{\mathrm{d}\tau} + \frac{\alpha-1}{t-\tau}\frac{\partial L}{\partial \dot{q}_s}\right) = vQ_s, \quad (s = 1, 2, \cdots, n). \tag{42}$$

Accordingly, Equation (37) is changed to

$$\begin{aligned}
&(1+\gamma)(t-\tau)^{\alpha-1}L^{\gamma}\left[\frac{\partial X^{(1)}(L)}{\partial q_s} - \frac{\mathrm{d}}{\mathrm{d}\tau}\frac{\partial X^{(1)}(L)}{\partial \dot{q}_s} - \frac{\gamma}{L}\frac{\mathrm{d}L}{\mathrm{d}\tau}\frac{\partial X^{(1)}(L)}{\partial \dot{q}_s} - \frac{\gamma}{L}\frac{\partial L}{\partial \dot{q}_s}\frac{\mathrm{d}X^{(1)}(L)}{\mathrm{d}\tau}\right. \\
&\quad \left. + \frac{\gamma}{L^2}\frac{\partial L}{\partial \dot{q}_s}\frac{\mathrm{d}L}{\mathrm{d}\tau}X^{(1)}(L) + \frac{\alpha-1}{t-\tau}\frac{\partial X^{(1)}(L)}{\partial \dot{q}_s}\right] = vX^{(1)}(Q_s), \quad (s = 1, 2, \cdots, n).
\end{aligned} \tag{43}$$

Substituting the formulae (11) into Equation (43), we get

$$\begin{aligned}
&(1+\gamma)(t-\tau)^{\alpha-1}v^m L^{\gamma}\left[\frac{\partial X_m^{(1)}(L)}{\partial q_s} - \frac{\mathrm{d}}{\mathrm{d}\tau}\frac{\partial X_m^{(1)}(L)}{\partial \dot{q}_s} - \frac{\gamma}{L}\frac{\mathrm{d}L}{\mathrm{d}\tau}\frac{\partial X_m^{(1)}(L)}{\partial \dot{q}_s} - \frac{\gamma}{L}\frac{\partial L}{\partial \dot{q}_s}\frac{\mathrm{d}X_m^{(1)}(L)}{\mathrm{d}\tau}\right. \\
&\quad \left. + \frac{\gamma}{L^2}\frac{\partial L}{\partial \dot{q}_s}\frac{\mathrm{d}L}{\mathrm{d}\tau}X_m^{(1)}(L) + \frac{\alpha-1}{t-\tau}\frac{\partial X_m^{(1)}(L)}{\partial \dot{q}_s}\right] = v^{m+1}X_m^{(1)}(Q_s), \quad (s = 1, 2, \cdots, n).
\end{aligned} \tag{44}$$

Therefore, we have

Theorem 4. *If the quasi-fractional dynamical system (34) is disturbed by small disturbance vQ_s, and there is a gauge function $G = G\big(\tau, q_k, \dot{q}_k\big)$ such that the structural equation*

$$\Big(\tfrac{1-\alpha}{t-\tau}\varsigma_0^m + \dot{\varsigma}_0^m\Big)X_m^{(1)}\big(L^{1+\gamma}\big)(t-\tau)^{\alpha-1} + X_m^{(1)}\Big[X_m^{(1)}\big(L^{1+\gamma}\big)\Big](t-\tau)^{\alpha-1} - X_{m-1}^{(1)}(Q_s)\big(\xi_s^{m-1} - \dot{q}_s\varsigma_0^{m-1}\big)$$
$$-\tfrac{\gamma}{L}Q_s X_{m-1}^{(1)}(L)\big(\xi_s^{m-1} - \dot{q}_s\varsigma_0^{m-1}\big) + \dot{G}^m = 0, \quad (s = 1, 2, \cdots, n;\ m = 0, 1, 2, \cdots), \tag{45}$$

holds, where $G = \sum\limits_{m=0}^{z} v^m G^m$ and $\varsigma_0^{-1} = \xi_s^{-1} = 0$, the Mei symmetry directly leads to the new adiabatic invariant

$$I_z = \sum_{m=0}^{z} v^m \left[(t-\tau)^{\alpha-1} X_m^{(1)}\big(L^{1+\gamma}\big)\varsigma_0^m + (t-\tau)^{\alpha-1}\frac{\partial}{\partial \dot{q}_s} X_m^{(1)}\big(L^{1+\gamma}\big)\big(\xi_s^m - \dot{q}_s\varsigma_0^m\big) + G^m \right]. \tag{46}$$

Proof. By using Equations (42), (44), and (45), we have

$$\begin{aligned}
\frac{\mathrm{d}I_z}{\mathrm{d}\tau} &= \sum_{m=0}^{z} v^m \Bigg\{ \tfrac{1-\alpha}{t-\tau}(t-\tau)^{\alpha-1}X_m^{(1)}\big(L^{1+\gamma}\big)\varsigma_0^m + (t-\tau)^{\alpha-1}\frac{\mathrm{d}X_m^{(1)}\big(L^{1+\gamma}\big)}{\mathrm{d}\tau}\varsigma_0^m \\
&\quad + (t-\tau)^{\alpha-1}X_m^{(1)}\big(L^{1+\gamma}\big)\dot{\varsigma}_0^m + \tfrac{1-\alpha}{t-\tau}(t-\tau)^{\alpha-1}\frac{\partial X_m^{(1)}\big(L^{1+\gamma}\big)}{\partial \dot{q}_s}\big(\xi_s^m - \dot{q}_s\varsigma_0^m\big) \\
&\quad + (t-\tau)^{\alpha-1}\frac{\mathrm{d}}{\mathrm{d}\tau}\frac{\partial X_m^{(1)}\big(L^{1+\gamma}\big)}{\partial \dot{q}_s}\big(\xi_s^m - \dot{q}_s\varsigma_0^m\big) + (t-\tau)^{\alpha-1}\frac{\partial X_m^{(1)}\big(L^{1+\gamma}\big)}{\partial \dot{q}_s}\big(\dot{\xi}_s^m - \dot{q}_s\dot{\varsigma}_0^m - \ddot{q}_s\varsigma_0^m\big) + \dot{G}^m \Bigg\} \\
&= \sum_{m=0}^{z} v^m \Bigg\{ \left[-\frac{\partial X_m^{(1)}(L)}{\partial q_s} + \frac{\mathrm{d}}{\mathrm{d}\tau}\frac{\partial X_m^{(1)}(L)}{\partial \dot{q}_s} + \tfrac{\gamma}{L}\frac{\mathrm{d}L}{\mathrm{d}\tau}\frac{\partial X_m^{(1)}(L)}{\partial \dot{q}_s} + \tfrac{\gamma}{L}\frac{\partial L}{\partial \dot{q}_s}\frac{\mathrm{d}X_m^{(1)}(L)}{\mathrm{d}\tau} \right. \\
&\quad \left. -\tfrac{\gamma}{L^2}\frac{\partial L}{\partial \dot{q}_s}\frac{\mathrm{d}L}{\mathrm{d}\tau}X_m^{(1)}(L) + \tfrac{1-\alpha}{t-\tau}\frac{\partial X_m^{(1)}(L)}{\partial \dot{q}_s} \right](1+\gamma)(t-\tau)^{\alpha-1}L^\gamma\big(\xi_s^m - \dot{q}_s\varsigma_0^m\big) \\
&\quad + \left(-\frac{\partial L}{\partial q_s} + \frac{\mathrm{d}}{\mathrm{d}\tau}\frac{\partial L}{\partial \dot{q}_s} + \tfrac{\gamma}{L}\frac{\partial L}{\partial \dot{q}_s}\frac{\mathrm{d}L}{\mathrm{d}\tau} + \tfrac{1-\alpha}{t-\tau}\frac{\partial L}{\partial \dot{q}_s} \right)(1+\gamma)(t-\tau)^{\alpha-1}X_m^{(1)}(L^\gamma)\big(\xi_s^m - \dot{q}_s\varsigma_0^m\big) \\
&\quad + \Big(\tfrac{1-\alpha}{t-\tau}\varsigma_0^m + \dot{\varsigma}_0^m \Big)X_m^{(1)}\big(L^{1+\gamma}\big)(t-\tau)^{\alpha-1} + X_m^{(1)}(L)X_m^{(1)}(L)(1+\gamma)\gamma(t-\tau)^{\alpha-1}L^{\gamma-1} \\
&\quad + X_m^{(1)}\Big[X_m^{(1)}(L)\Big](1+\gamma)(t-\tau)^{\alpha-1}L^\gamma + \dot{G}^m \Bigg\} \\
&= \sum_{m=0}^{z} v^m \Bigg\{ -v X_m^{(1)}(Q_s)\big(\xi_s^m - \dot{q}_s\varsigma_0^m\big) - v\tfrac{\gamma}{L}Q_s X_m^{(1)}(L)\big(\xi_s^m - \dot{q}_s\varsigma_0^m\big) \\
&\quad + X_{m-1}^{(1)}(Q_s)\big(\xi_s^{m-1} - \dot{q}_s\varsigma_0^{m-1}\big) + \tfrac{\gamma}{L}Q_s X_{m-1}^{(1)}(L)\big(\xi_s^{m-1} - \dot{q}_s\varsigma_0^{m-1}\big) \Bigg\} \\
&= -v^{z+1}\Big[X_z^{(1)}(Q_s) + \tfrac{\gamma}{L}Q_s X_z^{(1)}(L) \Big]\big(\xi_s^z - \dot{q}_s\varsigma_0^z\big).
\end{aligned} \tag{47}$$

According to the definition of adiabatic invariant [51], I_z is an adiabatic invariant of order z. So that ends the proof. □

Example 2. *Considering the nonconservative dynamical system, its action functional based on power-law Lagrangian is [50]*

$$A = \frac{1}{\Gamma(\alpha)}\int_{t_1}^{t_2}\Big[L^{1+\gamma}\big(\tau, q_s, \dot{q}_s\big)\Big](t-\tau)^{\alpha-1}\mathrm{d}\tau, \tag{48}$$

where $L = \dot{q} - q(\tau - t), \gamma = 1$.

Equation (34) gives

$$2(t-\tau)^{\alpha-1}\left[-\ddot{q} + \frac{\alpha-1}{t-\tau}\dot{q} + \big((\tau-t)^2 + \alpha\big)q \right] = 0. \tag{49}$$

By calculation, we have

$$X_0^{(1)}(L) = -q\varsigma_0^0 - (\tau-t)\xi_0^0 + \dot{\xi}_0^0 - \dot{q}\dot{\varsigma}_0^0, \tag{50}$$

$$X_0^{(1)}(L^2) = 2L\left[-q\varsigma_0^0 - (\tau - t)\xi_0^0 + \dot{\xi}_0^0 - \dot{q}\varsigma_0^0\right].\tag{51}$$

Let

$$\varsigma_0^0 = 0, \; \xi_0^0 = \exp\left[(t + \tau)^2/2\right],\tag{52}$$

then

$$X_0^{(1)}(L) = 0, \; X_0^{(1)}(L^2) = 0.\tag{53}$$

According to the criterion (37), the infinitesimals (52) correspond to Mei symmetry. Substituting (52) into Equation (38), we get

$$G^0 = 2\left[\dot{q} - q(\tau - t)\right](t - \tau)^{\alpha - 1}\exp\left[(\tau - t)^2/2\right].\tag{54}$$

From Theorem 3, we have

$$I_0 = 2\left[\dot{q} - q(\tau - t)\right](t - \tau)^{\alpha - 1}\exp\left[(\tau - t)^2/2\right] = \text{const.}\tag{55}$$

Let the small disturbance be

$$vQ = v\sin\tau\exp\left[-(\tau - t)^2/2\right].\tag{56}$$

The differential equation of the disturbed motion is

$$2(t - \tau)^{\alpha - 1}\left[-\ddot{q} + \frac{\alpha - 1}{t - \tau}\dot{q} + \left((\tau - t)^2 + \alpha\right)q\right] = v\sin\tau\exp\left[-(\tau - t)^2/2\right].\tag{57}$$

Take

$$\varsigma_0^1 = 0, \; \xi_0^1 = \exp\left[(t + \tau)^2/2\right],\tag{58}$$

then it is easy to verify

$$X_1^{(1)}(L) = 0, \; X_1^{(1)}(L^2) = 0, \; X_0^{(1)}(Q) = X_1^{(1)}(Q) = 0.\tag{59}$$

According to the criterion (44), the infinitesimals (58) correspond to Mei symmetry. Substituting (58) into Equation (45), we have

$$G^1 = 2\left[\dot{q} - q(\tau - t)\right](t - \tau)^{\alpha - 1}\exp\left[(\tau - t)^2/2\right] - v\cos\tau.\tag{60}$$

By Theorem 4, we obtain

$$\begin{aligned}
I_1 &= 2\left[\dot{q} - q(\tau - t)\right](t - \tau)^{\alpha - 1}\exp\left[(\tau - t)^2/2\right]\\
&+ v\left\{2\left[\dot{q} - q(\tau - t)\right](t - \tau)^{\alpha - 1}\exp\left[(\tau - t)^2/2\right] - v\cos\tau\right\}.
\end{aligned}\tag{61}$$

Formula (61) is an adiabatic invariant led by Mei symmetry.

4. Conclusions

Symmetry is closely related to invariants, and it is of great significance to find the invariants of complex system dynamics. First, even if the equations of motion are difficult to solve, the existence of some conserved quantity makes it possible to understand the local physical state or dynamical behavior of the system. Secondly, we can reduce the differential equations of motion by using conserved quantities. Thirdly, we can study the motion stability of complex dynamical systems by using conserved quantities. Based on the quasi-fractional dynamical model proposed by El-Nabulsi according to the Riemann–Liouville definition of fractional integral, we studied Mei symmetry and its corresponding invariants of quasi-fractional dynamics system whose action functional is composed of non-standard Lagrangians. The main results of this paper are its four theorems. In this paper, we provided a method to study nonlinear or non-conservative dynamics and obtained new conserved quantities and new adiabatic invariants, and the results are expected to be generalized or applied to the dynamics of constrained systems, such as those of nonholonomic systems.

Author Contributions: The authors equally contributed to this research work.

Funding: This work was supported by the National Natural Science Foundation of China (grant nos. 11,572,212 and 11,272,227).

Conflicts of Interest: The authors declare no conflict of interest.

Appendix A. Derivation of the Euler–Lagrange Equations for Quasi-Fractional Dynamical System with Exponential Lagrangians

Consider a nonlinear dynamical system whose configuration is determined by n generalized coordinates $q_s (s = 1, 2, \cdots, n)$, its action functional based on exponential Lagrangian is

$$S = \frac{1}{\Gamma(\alpha)} \int_{t_1}^{t_2} \exp\left[L\left(\tau, q_s, \dot{q}_s\right)\right](t - \tau)^{\alpha-1} \mathrm{d}\tau. \tag{A1}$$

where $L = L\left(\tau, q_s, \dot{q}_s\right)$ is the standard Lagrangian, $0 < \alpha \leq 1$, τ is the intrinsic time, t is the observer time, and τ is not equal to t.

The isochronous variational principle

$$\delta S = 0, \tag{A2}$$

which satisfies the following commutation relation

$$\mathrm{d}\delta q_s = \delta \mathrm{d}q_s, \ (s = 1, 2, \cdots, n), \tag{A3}$$

and given boundary condition

$$\delta q_s\big|_{t=t_1} = \delta q_s\big|_{t=t_2} = 0, \ (s = 1, 2, \cdots, n) \tag{A4}$$

can be called the Hamilton principle of the quasi-fractional dynamical system with exponential Lagrangians.

Expanding the Hamilton principle (A2), we have

$$\begin{aligned}
0 = \delta S &= \tfrac{1}{\Gamma(\alpha)} \int_{t_1}^{t_2} \delta\left[\exp L(t - \tau)^{\alpha-1}\right] \mathrm{d}\tau \\
&= \tfrac{1}{\Gamma(\alpha)} \int_{t_1}^{t_2} (t - \tau)^{\alpha-1} \exp L\left(\tfrac{\partial L}{\partial q_s}\delta q_s + \tfrac{\partial L}{\partial \dot{q}_s}\delta \dot{q}_s\right)\mathrm{d}\tau
\end{aligned} \tag{A5}$$

Due to

$$\begin{aligned}
\int_{t_1}^{t_2} (t - \tau)^{\alpha-1} \exp L \tfrac{\partial L}{\partial \dot{q}_s}\delta \dot{q}_s \mathrm{d}\tau &= \left[(t - \tau)^{\alpha-1} \exp L \tfrac{\partial L}{\partial \dot{q}_s}\delta q_s\right]\Big|_{t_1}^{t_2} \\
&\quad - \int_{t_1}^{t_2} (t - \tau)^{\alpha-1} \exp L\left(-\tfrac{\alpha-1}{t-\tau}\tfrac{\partial L}{\partial \dot{q}_s} + \tfrac{\mathrm{d}L}{\mathrm{d}\tau}\tfrac{\partial L}{\partial \dot{q}_s} + \tfrac{\mathrm{d}}{\mathrm{d}\tau}\tfrac{\partial L}{\partial \dot{q}_s}\right)\delta q_s \mathrm{d}\tau \\
&= -\int_{t_1}^{t_2} (t - \tau)^{\alpha-1} \exp L\left(-\tfrac{\alpha-1}{t-\tau}\tfrac{\partial L}{\partial \dot{q}_s} + \tfrac{\mathrm{d}L}{\mathrm{d}\tau}\tfrac{\partial L}{\partial \dot{q}_s} + \tfrac{\mathrm{d}}{\mathrm{d}\tau}\tfrac{\partial L}{\partial \dot{q}_s}\right)\delta q_s \mathrm{d}\tau.
\end{aligned} \tag{A6}$$

Substituting the formula (A6) into Equation (A5), we have

$$\frac{1}{\Gamma(\alpha)} \int_{t_1}^{t_2} (t - \tau)^{\alpha-1} \exp L\left(\frac{\partial L}{\partial q_s} - \frac{\mathrm{d}}{\mathrm{d}\tau}\frac{\partial L}{\partial \dot{q}_s} - \frac{\partial L}{\partial \dot{q}_s}\frac{\mathrm{d}L}{\mathrm{d}\tau} + \frac{\alpha-1}{t-\tau}\frac{\partial L}{\partial \dot{q}_s}\right)\delta q_s \mathrm{d}\tau = 0. \tag{A7}$$

Because of the arbitrariness of the interval $[t_1, t_2]$ and the independence of δq_s $(s = 1, 2, \cdots, n)$, using the fundamental lemma [23] of the calculus of variations, we get

$$(t - \tau)^{\alpha-1} \exp L\left(\frac{\partial L}{\partial q_s} - \frac{\mathrm{d}}{\mathrm{d}\tau}\frac{\partial L}{\partial \dot{q}_s} - \frac{\partial L}{\partial \dot{q}_s}\frac{\mathrm{d}L}{\mathrm{d}\tau} + \frac{\alpha - 1}{t - \tau}\frac{\partial L}{\partial \dot{q}_s}\right) = 0, \ (s = 1, 2, \cdots, n). \tag{A8}$$

Equation (A8) can be called the Euler–Lagrange equations for quasi-fractional dynamical system with exponential Lagrangians.

Appendix B. Derivation of the Euler–Lagrange Equations for Quasi-Fractional Dynamical System with Power-Law Lagrangians

Consider a nonlinear dynamical system whose configuration is determined by n generalized coordinates $q_s(s = 1, 2, \cdots, n)$, its action functional based on power-law Lagrangian is

$$A = \frac{1}{\Gamma(\alpha)} \int_{t_1}^{t_2} \left[L^{1+\gamma}\left(\tau, q_s, \dot{q}_s\right)\right](t - \tau)^{\alpha-1}\mathrm{d}\tau \tag{A9}$$

where $L = L\left(\tau, q_s, \dot{q}_s\right)$ is the standard Lagrangian, γ is not equal to -1, $0 < \alpha \leq 1$, τ is the intrinsic time, t is the observer time, and τ is not equal to t.

The isochronous variational principle

$$\delta A = 0, \tag{A10}$$

which satisfies the following commutation relation

$$\mathrm{d}\delta q_s = \delta \mathrm{d}q_s, \ (s = 1, 2, \cdots, n), \tag{A11}$$

and given boundary condition

$$\delta q_s\big|_{t=t_1} = \delta q_s\big|_{t=t_2} = 0, \ (s = 1, 2, \cdots, n) \tag{A12}$$

can be called the Hamilton principle of the quasi-fractional dynamical system with power-law Lagrangians.

Expanding the Hamilton principle (A10), we have

$$\begin{aligned} 0 = \delta A &= \frac{1}{\Gamma(\alpha)} \int_{t_1}^{t_2} \delta\left[L^{1+\gamma}(t - \tau)^{\alpha-1}\right]\mathrm{d}\tau \\ &= \frac{1}{\Gamma(\alpha)} \int_{t_1}^{t_2} (1 + \gamma)(t - \tau)^{\alpha-1}L^\gamma\left(\frac{\partial L}{\partial q_s}\delta q_s + \frac{\partial L}{\partial \dot{q}_s}\delta \dot{q}_s\right)\mathrm{d}\tau \end{aligned} \tag{A13}$$

Due to

$$\begin{aligned} \int_{t_1}^{t_2} (t - \tau)^{\alpha-1}L^\gamma\frac{\partial L}{\partial \dot{q}_s}\delta \dot{q}_s\mathrm{d}\tau &= \left[(t - \tau)^{\alpha-1}L^\gamma\frac{\partial L}{\partial \dot{q}_s}\delta q_s\right]\Big|_{t_1}^{t_2} \\ &\quad - \int_{t_1}^{t_2} (t - \tau)^{\alpha-1}L^\gamma\left(-\frac{\alpha-1}{t-\tau}\frac{\partial L}{\partial \dot{q}_s} + \frac{\gamma}{L}\frac{\mathrm{d}L}{\mathrm{d}\tau}\frac{\partial L}{\partial \dot{q}_s} + \frac{\mathrm{d}}{\mathrm{d}\tau}\frac{\partial L}{\partial \dot{q}_s}\right)\delta q_s\mathrm{d}\tau \\ &= -\int_{t_1}^{t_2} (t - \tau)^{\alpha-1}L^\gamma\left(-\frac{\alpha-1}{t-\tau}\frac{\partial L}{\partial \dot{q}_s} + \frac{\gamma}{L}\frac{\mathrm{d}L}{\mathrm{d}\tau}\frac{\partial L}{\partial \dot{q}_s} + \frac{\mathrm{d}}{\mathrm{d}\tau}\frac{\partial L}{\partial \dot{q}_s}\right)\delta q_s\mathrm{d}\tau. \end{aligned} \tag{A14}$$

Substituting the formula (A14) into Equation (A13), we have

$$\frac{1}{\Gamma(\alpha)} \int_{t_1}^{t_2} (1 + \gamma)(t - \tau)^{\alpha-1}L^\gamma\left(\frac{\partial L}{\partial q_s} - \frac{\mathrm{d}}{\mathrm{d}\tau}\frac{\partial L}{\partial \dot{q}_s} - \frac{\gamma}{L}\frac{\mathrm{d}L}{\mathrm{d}\tau}\frac{\partial L}{\partial \dot{q}_s} + \frac{\alpha - 1}{t - \tau}\frac{\partial L}{\partial \dot{q}_s}\right)\delta q_s\mathrm{d}\tau = 0. \tag{A15}$$

Because of the arbitrariness of the interval $[t_1, t_2]$ and the independence of δq_s $(s = 1, 2, \cdots, n)$, using the fundamental lemma [23] of the calculus of variations, we get

$$(1 + \gamma)(t - \tau)^{\alpha - 1} L^\gamma \left(\frac{\partial L}{\partial q_s} - \frac{\mathrm{d}}{\mathrm{d}\tau} \frac{\partial L}{\partial \dot{q}_s} - \frac{\gamma}{L} \frac{\partial L}{\partial \dot{q}_s} \frac{\mathrm{d}L}{\mathrm{d}\tau} + \frac{\alpha - 1}{t - \tau} \frac{\partial L}{\partial \dot{q}_s} \right) = 0, \ (s = 1, 2, \cdots, n). \tag{A16}$$

Equation (A16) can be called the Euler–Lagrange equations for quasi-fractional dynamical system with power-law Lagrangians. Equation (A16) is consistent with the results given in [50].

References

1. Noether, A.E. Invariante Variationsprobleme. Nachrichten von der Gesellschaft der Wissenschaften zu Göttingen. *Math. Phys. Kl.* **1918**, *2*, 235–257.
2. Lutzky, M. Dynamical symmetries and conserved quantities. *J. Phys. A Math. Gen.* **1979**, *12*, 973–981. [CrossRef]
3. Bluman, G.W.; Anco, S.C. *Symmetry and Integration Methods for Differential Equations*; Springer: New York, NY, USA, 2002.
4. Mei, F.X. Form invariance of Lagrange system. *J. Beijing Inst. Technol.* **2000**, *9*, 120–124.
5. Mei, F.X. *Symmetries and Conserved Quantities of Constrained Mechanical Systems*; Beijing Institute of Technology Press: Beijing, China, 2004.
6. Hojman, S.A. A new conservation law constructed without using either Lagrangians or Hamiltonians. *J. Phys. A Math. Gen.* **1992**, *25*, L291–L295. [CrossRef]
7. Ma, W.X. Conservation laws of discrete evolution equations by symmetries and adjoint symmetries. *Symmetry* **2015**, *7*, 714–725. [CrossRef]
8. Ma, W.X. Conservation laws by symmetries and adjoint symmetries. *Discret. Cont. Dyn. S* **2018**, *11*, 707–721. [CrossRef]
9. Mei, F.X. Advances in the symmetries and conserved quantities of classical constrained systems. *Adv. Mech.* **2009**, *39*, 37–43.
10. Galiullin, A.S.; Gafarov, G.G.; Malaishka, R.P.; Khwan, A.M. *Analytical Dynamics of Helmholtz, Birkhoff and Nambu Systems*; UFN: Moscow, Russia, 1997.
11. Mei, F.X. Lie symmetries and conserved quantities of constrained mechanical systems. *Acta Mech.* **2000**, *141*, 135–148. [CrossRef]
12. Zhang, Y.; Zhai, X.H. Noether symmetries and conserved quantities for fractional Birkhoffian systems. *Nonlinear Dyn.* **2015**, *81*, 469–480. [CrossRef]
13. Zhai, X.H.; Zhang, Y. Lie symmetry analysis on time scales and its application on mechanical systems. *J. Vib. Control* **2019**, *25*, 581–592. [CrossRef]
14. Jia, L.Q.; Wang, X.X.; Zhang, M.L.; Han, Y.L. Special Mei symmetry and approximate conserved quantity of Appell equations for a weakly nonholonomic system. *Nonlinear Dyn.* **2012**, *69*, 1807–1812. [CrossRef]
15. Zhang, Y. Noether's theorem for a time-delayed Birkhoffian system of Herglotz type. *Int. J. Non-Linear Mech.* **2018**, *101*, 36–43. [CrossRef]
16. Djukić, D.S. Adiabatic invariants for dynamical systems with one degree of freedom. *Int. J. Non-Linear Mech.* **1981**, *16*, 489–498. [CrossRef]
17. Jiang, W.A.; Luo, S.K. A new type of non-Noether exact invariants and adiabatic invariants of generalized Hamiltonian systems. *Nonlinear Dyn.* **2012**, *67*, 475–482. [CrossRef]
18. Song, C.J.; Zhang, Y. Conserved quantities and adiabatic invariants for fractional generalized Birkhoffian systems. *Int. J. Non-Linear Mech.* **2017**, *90*, 32–38. [CrossRef]
19. Yang, M.J.; Luo, S.K. Fractional symmetrical perturbation method of finding adiabatic invariants of disturbed dynamical systems. *Int. J. Non-Linear Mech.* **2018**, *101*, 16–25. [CrossRef]
20. Zhang, Y.; Wang, X.P. Lie symmetry perturbation and adiabatic invariants for dynamical system with non-standard Lagrangians. *Int. J. Non-Linear Mech.* **2018**, *105*, 165–172. [CrossRef]
21. Luo, S.K.; Yang, M.J.; Zhang, X.T.; Dai, Y. Basic theory of fractional Mei symmetrical perturbation and its application. *Acta Mech.* **2018**, *229*, 1833–1848. [CrossRef]

22. Zhang, Y.; Zhai, X.H. Perturbation to Lie symmetry and adiabatic invariants for Birkhoffian systems on time scales. *Commun. Nonlinear Sci. Numer. Simulat.* **2019**, *75*, 251–261. [CrossRef]
23. Arnold, V.I. *Mathematical Methods of Classical Mechanics*; Springer: New York, NY, USA, 1978.
24. Alekseev, A.I.; Arbuzov, B.A. Classical Yang-Mills field theory with nonstandard Lagrangians. *Theor. Math. Phys.* **1984**, *59*, 372–378. [CrossRef]
25. Musielak, Z.E. Standard and non-standard Lagrangians for dissipative dynamical systems with variable coefficients. *J. Phys. A Math. Theor.* **2008**, *41*, 055205. [CrossRef]
26. El-Nabulsi, R.A. Nonlinear dynamics with nonstandard Lagrangians. *Qual. Theory Dyn. Syst.* **2012**, *12*, 273–291. [CrossRef]
27. El-Nabulsi, R.A. Non-Standard non-local-in-time Lagrangians in classical mechanics. *Qual. Theory Dyn. Syst.* **2014**, *13*, 149–160. [CrossRef]
28. El-Nabulsi, R.A. Fractional oscillators from non-standard Lagrangians and time-dependent fractional exponent. *Comput. Appl. Math.* **2014**, *33*, 163–179. [CrossRef]
29. Dimitrijevic, D.D.; Milosevic, M. About non-standard Lagrangians in cosmology. *AIP Conf. Proc.* **2012**, *1472*, 41.
30. Zhang, Y.; Zhou, X.S. Noether theorem and its inverse for nonlinear dynamical systems with non-standard Lagrangians. *Nonlinear Dyn.* **2016**, *84*, 1867–1876. [CrossRef]
31. Song, J.; Zhang, Y. Noether symmetry and conserved quantity for dynamical system with non-standard Lagrangians on time scales. *Chin. Phys. B* **2017**, *26*, 201–209. [CrossRef]
32. Song, J.; Zhang, Y. Noether's theorems for dynamical systems of two kinds of non-standard Hamiltonians. *Acta Mech.* **2018**, *229*, 285–297. [CrossRef]
33. Fiori, S. Extended Hamiltonian learning on Riemannian manifolds: Theoretical aspects. *IEEE T. Neur. Net. Lear.* **2011**, *22*, 687–700. [CrossRef]
34. Fiori, S. Extended Hamiltonian learning on Riemannian manifolds: Numerical aspects. *IEEE T. Neur. Net. Lear.* **2012**, *23*, 7–21. [CrossRef]
35. Oldham, K.B.; Spanier, J. *The Fractional Calculus*; Academic Press: San Diego, CA, USA, 1974.
36. Podlubny, I. *Fractional Differential Equations*; Academic Press: San Diego, CA, USA, 1999.
37. Kilbas, A.A.; Srivastava, H.M.; Trujillo, J.J. *Theory and Applications of Fractional Differential Equations*; Elsevier BV: Amsterdam, The Netherland, 2006.
38. Riewe, F. Nonconservative Lagrangian and Hamiltonian mechanics. *Phys. Rev. E* **1996**, *53*, 1890–1899. [CrossRef]
39. Riewe, F. Mechanics with fractional derivatives. *Phys. Rev. E* **1997**, *55*, 3581–3592. [CrossRef]
40. Agrawal, O.P. Formulation of Euler-Lagrange equations for fractional variational problems. *J. Math. Anal. Appl.* **2002**, *272*, 368–379. [CrossRef]
41. Baleanu, D.; Trujillo, J.I. A new method of finding the fractional Euler-Lagrange and Hamilton equations within Caputo fractional derivatives. *Commun. Nonlinear Sci. Numer. Simulat.* **2010**, *15*, 1111–1115. [CrossRef]
42. Atanacković, T.M.; Konjik, S.; Pilipović, S.; Simić, S. Variational problems with fractional derivatives: Invariance conditions and Noether's theorem. *Nonlinear Anal. Theory* **2009**, *71*, 1504–1517. [CrossRef]
43. Malinowska, A.B.; Torres, D.F.M. *Introduction to the Fractional Calculus of Variations*; Imperial College Press: London, UK, 2012.
44. Li, M. Three classes of fractional oscillators. *Symmetry* **2018**, *10*, 40. [CrossRef]
45. Zhai, X.H.; Zhang, Y. Noether symmetries and conserved quantities for fractional Birkhoffian systems with time delay. *Commun. Nonlinear Sci. Numer. Simulat.* **2016**, *36*, 81–97. [CrossRef]
46. Yan, B.; Zhang, Y. Noethe's theorem for fractional Birkhoffian systems of variable order. *Acta Mech.* **2016**, *227*, 2439–2449. [CrossRef]
47. Meng, W.; Zeng, B.; Li, S.L. A novel fractional-order grey prediction model and its modeling error analysis. *Information* **2019**, *10*, 167. [CrossRef]
48. El-Nabulsi, R.A. A fractional approach to nonconservative Lagrangian dynamical systems. *Fizika A* **2005**, *14*, 289–298.
49. El-Nabulsi, R.A.; Torres, D.F.M. Fractional action-like variational problems. *J. Math. Phys.* **2008**, *49*, 053521. [CrossRef]

50. El-Nabulsi, R.A. Non-standard fractional Lagrangians. *Nonlinear Dyn.* **2013**, *74*, 381–394. [CrossRef]
51. Zhao, Y.Y.; Mei, F.X. *Symmetries and Invariants of Mechanical Systems*; Science Press: Beijing, China, 1999.

Article

Multi-Agent Reinforcement Learning Using Linear Fuzzy Model Applied to Cooperative Mobile Robots

David Luviano-Cruz [1,*], Francesco Garcia-Luna [1], Luis Pérez-Domínguez [1] and S. K. Gadi [2]

[1] Department of industrial engineering and manufacturing, Autonomous University of Ciudad Juarez, Ciudad Juarez 32310, Mexico; francesco.garcia@uacj.mx (F.G.-L.); luis.dominguez@uacj.mx (L.P.-D.)
[2] Faculty of mechanical and electrical engineering, Autonomous University of Coahuila, Torreon 27276, Mexico; research@skgadi.com
* Correspondence: david.luviano@uacj.mx; Tel.: +52-688-2100

Received: 5 September 2018; Accepted: 30 September 2018; Published: 3 October 2018

Abstract: A multi-agent system (MAS) is suitable for addressing tasks in a variety of domains without any programmed behaviors, which makes it ideal for the problems associated with the mobile robots. Reinforcement learning (RL) is a successful approach used in the MASs to acquire new behaviors; most of these select exact Q-values in small discrete state space and action space. This article presents a joint Q-function linearly fuzzified for a MAS' continuous state space, which overcomes the dimensionality problem. Also, this article gives a proof for the convergence and existence of the solution proposed by the algorithm presented. This article also discusses the numerical simulations and experimental results that were carried out to validate the proposed algorithm.

Keywords: multi-agent system (MAS); reinforcement learning (RL); mobile robots; function approximation

---

1. Introduction

Multi-agent systems (MASs) are finding application in a variety of fields where pre-programmed behaviors are not a suitable way to tackle the problems that arise. These fields include robotics, distributed control, resource management, collaborative decision making, data mining [1]. A MAS includes several intelligent agents in an environment, where each one has its independent behavior and should coordinate with the others [2].

MASs could emerge as an alternative way to analyze and represent the systems with centralized control, where several intelligent agents perceive and modify an environment through sensors and actuators respectively. At the same time, these agents can also learn new behaviors to adapt themselves to the new tasks and the goals in an environment [3].

One of the fields where multi-agent systems have emerged are mobile robots, most approaches are based on low level control systems, in [4] a visibility binary tree algorithm is used to generate the mobile robot trajectories. This type of approach is based on the complete knowledge of the dynamics of the robotic system. In this article, we offer a proposal based on reinforcement learning, which will result in high-level control actions.

Reinforcement learning (RL) is one of the most popular methods for learning in a MAS. The objective of a Multi-agent reinforcement learning (MARL) is to maximize a numerical reward; so that, the agents can interact with the environment and modify it [5]. At each learning step, these agents choose an action, which drives the environment to a new state [6]. The Reward function assesses the grade of this state transition [7]. In the RL the agents are not told which tasks should be executed instead they must explore which actions have the best reward. Hence, the RL feedback is less informative than a supervised learning method [8].

MASs are affected by the curse of dimensionality, which is a term given to suggest that the computational and memory requirements increase as the number of states or agents increase in an environment. Most approaches require an exact representation of the state-action pair values in the form of lookup tables, making the solution intractable, hence the application of these methods is reduced to small or discrete tasks [9]. In the real-life applications, the state variables can be a selected from of a large number of possible values or even from the continuous values; so, the problem is manageable if value functions are approximated [10].

Some MARL algorithms have been proposed to deal with this problem using neural networks by making generalizations from a Q-table [11], applying function approximation for discrete and large state-action space [12], applying vector quantization for continuous state and actions [13], using experience replay for MAS [14], using Q-learning and normalized Gaussian network as approximators [15], predictions in systems with heterogeneous agents [16]. In [17] a couple of neural networks is used to represent the value function and the controller, however, the proposed strategy is based on a sufficient exploration, which is a function of the size of the training data of the neural networks. Inverse neural networks have also been proposed to approximate the actions policy, which uses an initial policy of actions to be refined through reinforcement learning [18].

This article presents an approach for MARL in a cooperative problem. It is a modified version of the Q-learning algorithm proposed in [19] which uses a linear fuzzy approximator of the joint Q-function for a continuous state space. An implicit form of coordination is implemented to solve the coordination problem. An experiment is conducted on two robots performing the task of solving a coordination problem to verify the proposed algorithm. In addition to that, two theorems are presented to ensure the convergence of the proposed algorithm.

2. MARL with Linear Fuzzy Parameterization

2.1. Single Agent Case

In a reinforcement learning (RL) for a single agent case, let us define: x_k as the current state of the environment at the learning step k, and u_k as the action taken by the agent in x_k.

The reward or the numerical feedback, r_k, reflects how good was the previous action, u_{k-1}, in the state x_{k-1}. The single agent RL problem is a Markov decision process (MDP). MDP for a deterministic case is:

$$f : X \times U \rightarrow X$$
$$\rho : X \times U \rightarrow \mathbb{R} \tag{1}$$

where X is the state space, U is the actions space, f is the state transition function which can be known or unknown, and ρ is the reward scalar function.

The action policy is used to describe the agent's behavior, which specifies the way in which the agent chooses the action from a state. If the action policy, $h = X \rightarrow U$, does not change over time it is considered stationary [20].

The final goal is to find an action policy h such that the long-term return R is maximized:

$$R^h = E \left\{ \sum_{k=0}^{\infty} \gamma^k r_{k+1} \,\middle|\, x_0 = x, \pi \right\} \tag{2}$$

where $\gamma \in [0, 1)$ is the discount factor. The policy, h, is obtained from the state-action value function, called Q-function.

The Q-function

$$Q^h : X \times U \rightarrow \mathbb{R} \tag{3}$$

gives a expected return from the policy, h,

$$Q^h(x, u) = E\left\{ \sum_{k=0}^{\infty} \gamma^k r_{k+1} \,\middle|\, x_0 = x, u_0 = u, h \right\} \tag{4}$$

The optimal Q-function Q^* is defined as:

$$Q^*(x, u) = \max_h Q^h(x, u) \tag{5}$$

Once Q^* is available, the optimal action policy is obtained by:

$$h^*(x) = \arg\max_u Q^*(x, u) \tag{6}$$

2.2. Multi Agent System Case

In a Multi-agent case, there is some number of heterogeneous agents with their own set of actions and tasks in an environment. A stochastic game model describes this behavior in which the action performed at any state is a combination of the actions by each agent [21].

The deterministic stochastic game's model is a tuple $(X, U_1, U_2, ..., U_n, f, \rho_1, \rho_2, ..., \rho_n)$, where n is the number of the agents in the environment, X is the state of the environment, U_i $i = 1, 2, ..., n$ are the sets of actions available to each agent and the joint action set $U = U_1 \times U_2 \times ... \times U_n$. The reward functions $\rho_i : X \times U \to R$, $i = 1, 2, ..., n$ and the state transition function is $f : X \times U \to X$.

The joint action $\mathbf{u}_k = \left[u_{1,k}^T, u_{2,k}^T, ..., u_{n,k}^T \right]^T$, $\mathbf{u}_k \in U$, $u_i \in U_i$ taken in the state x_k, changes the state to $x_{k+1} = f(x_k, \mathbf{u}_k)$. A numerical value for the reward is calculated as $r_{i,k+1} = \rho(x_k, \mathbf{u}_k)$ for each joint action $\mathbf{u}_k$. The actions are taken according to each agent's own policy $h_i : X \to U_i$, where all of them form the joint policy $\mathbf{h}$. Similar to a single agent case, the state space and actions space can be continuous or discrete.

The long term reward R depends on the joint policy $R_i^{\mathbf{h}}(x) = \sum_{k=0}^{\infty} \gamma^k r_{i,k+1}$ due to the numerical feedback r of each agent depends on the joint action $\mathbf{u}_k$. Thereby the Q-function of each agent relies on the joint action and the joint policy, $Q_i^h = X \times U \to R$, with $Q_i^h(x, u) = E\left[\sum_{k=0}^{\infty} \gamma^k r_{i,k+1} \,|\, x_0 = x, \mathbf{u}_0 = \mathbf{u}, \mathbf{h} \right]$.

Each agent could have its own goals, however, in this article the agents seek a common goal, i.e., the task is fully cooperative. In this way the numerical feedback or reward for any state is the same for all agents $\rho_1 = \rho_2 = ... = \rho_n$, therefore the reward scalar functions and returns are the same for all the agents, $R_1^{\mathbf{h}} = R_2^{\mathbf{h}} = ... = R_n^{\mathbf{h}}$. Hence the agents have the same goal which is maximize the common long term performance (or return).

Determining an optimal joint policy $\mathbf{h}^*$ in Multi-agent systems is the equilibrium selection problem [22]. Although establishing an equilibrium is a difficult problem, the structure of the cooperative settings make this problem manageable. Assuming that the agents know the structure of the game in the form of the transition function f and the reward function ρ_i makes the searching of the equilibrium point more tractable.

In a fully cooperative stochastic game, if the transition function f and the reward function ρ for each agent is known, the objective can be accomplished by learning the optimal joint-action values Q^* through Bellman optimal equation: $Q(x_k, \mathbf{u}_k) = \rho(x_k, \mathbf{u}_k) + \gamma \max_j Q(f(x_k, \mathbf{u}_k), \mathbf{u}_j)$ and then using a greedy policy [23]. Once Q^* is available, a policy h is:

$$h_i^*(x) = \arg\max_{u_i} \max_{u_1, u_2, ..., u_n} Q^*(x, \mathbf{u}) \tag{7}$$

When several joint actions are optimal the agents could choose different actions and degrade the performance of the search for a common goal. This problem can be solved by: the coordination free methods assume that the optimal join action are unique across learning a common Q-function [24], the coordination-based methods use the coordination graphs through a decomposition of the global Q-function in local Q-functions [25], or the implicit coordination methods assume the agents learn to choose one joint action by chance and then discard the others [26].

2.3. Mapping the Joint Q-Function for MAS

In a deterministic case, the Q-function, Q, is

$$\begin{aligned} Q_{k+1} &= H(Q_k) \\ H(Q)(x, \mathbf{u}) &= \rho(x, \mathbf{u}) + \gamma \max_j Q\left(f(x, \mathbf{u}), \mathbf{u}_j\right) \end{aligned} \tag{8}$$

with an arbitrary initial value for Q. The iterations (8) attracts the Q to a unique fixed point at [27]

$$Q^* = H(Q^*).$$

In [28] is shown that the mapping $H : \mathbf{Q} \to \mathbf{Q}$ is a contraction with factor $\alpha < 1$. For any pair of Q-function Q_1 and Q_2,

$$\left\| H(Q_1) - H(Q_2) \right\| \le \alpha \left\| Q_1 - Q_2 \right\| \tag{9}$$

and then H has a unique fixed point. Q^* is a fixed point of $H : Q^* = H(Q^*)$, and the iteration converges to Q^* as $k \to \infty$. An optimal policy $h_i^*(x)$ can be calculated from Q^* using (8). To perform the former iteration, we need a model of the task in the form of the transition function f and reward function ρ_i.

This kind of method, based on the Bellman optimality equation, need saving and updating the Q-values for each state-joint action stage. In this way, only tasks with finite discrete set of state and actions are generally treated. The dimensionality problems occur by the growth of the number of agents involved in the task [29], thus this generates an increment on the computational complexity [30].

In the case where the state space or actions space are continuous or even discrete with a great number of variables, the Q-functions must be depicted in an approximated form [31]. Because an exact representation of the Q-function could be impractical or intractable, therefore, we propose an approximate linear fuzzy representation of the joint Q-function through a vector ϕ.

2.4. Linear Fuzzy Parameterization of the Joint Q-Function

In general, if there is no prior awareness about the Q-function, the only form to have an exact representation is saving distinct Q-values for each state-action couple. If the state space is continuous, the exact representation of the Q-function would need take an infinite number of state-action values. For this reason, the only practical way to overcome this situation is using an approximate representation of the joint Q-function.

In this section, we present a parameterized version of the Q-function through a linear fuzzy approximator which consist in a vector $\phi \in R^z$, this vector relies in a fuzzy partition of the continuous state space. The principal advantage of this proposal is that we only need to save the state-action pair Q-value of the center of every membership function.

There are N fuzzy sets, which are depicted by a membership function:

$$\mu_d(x) = X \to [0,1] \quad d = 1, 2, .., N \tag{10}$$

where $\mu_d(x)$ describe the degree of membership of the state x to the fuzzy set d, this membership functions can be looked as basis functions or features [32]. The number of membership functions increase with the size of the state space, the number of the agents and with the degree of resolution sought for the vector ϕ.

Triangular shapes of fuzzy partitions are used in this paper since they have their maximum value in a single point, namely, for every d exist a unique x_d (the core of the membership function) such that

$\mu_d(x_d) > \mu_d(x) \; \forall x \neq x_d$. Since all the others membership functions take zero values in x_d, $\mu_{\hat{d}}(x_d) = 0$ for $\forall d \neq \hat{d}$, we assume that $\mu_d(x_d) = 1$, which mean that the membership functions are normal. Another kind of Membership Functions shape could diverge when they have too big values in x_d [33].

We have a number N of triangular membership functions for each state variable $x_e = 1, 2, ..., E$, with $\dim(X) = E$. A pyramid shape E dimensional of membership functions will be the consequence of the product of each single dimensional membership function in the fuzzy partition of the state space.

We assume that the action space is discrete for all the agents and they have the same number of actions available:

$$U_i = \left\{ u_{ij} \, | i = 1, 2, .., n \quad j = 1, 2, ..M \right\} \tag{11}$$

The parameter vector ϕ is composed by $z = nNM$ elements to be stored, the membership function-action pair (μ_d, u_{ij}) for each agent correspond an element of the parameter vector $\phi_{i,d,j}$. The approximator's elements $\phi_{i,d,j}$ are organized in a preliminary way using n different matrices with size $N \times M$, filling the first M columns with the N elements available. The elements of the n matrix are allocated in a vector arrangement ϕ column by column for the first agent, then follow with the next agent's columns until completing n agents.

Denoting the scalar indexes $[i, d, j]$ of ϕ by:

$$[i, d, j] = [d + (j - 1) N] + (i - 1) N \times M \tag{12}$$

where $i = 1, 2, ..., n$ means the number of the analyzed agent, $d = 1, 2, ..., N$ is the number of fuzzy partitions for each state variable $x_e = 1, 2, ..., E$, with $\dim(X) = E$ and $j = 1, 2, ..., M$ $\dim(U_i) = M$. In this way we denote the indexes of the parameter approximator by $\phi_{[i,d,j]}$, which means the approximate Q-value for the d membership function, performing the action j available for the agent i.

The state x is taken as input by the fuzzy rule base and produces M outputs for each agent, which are the corresponding Q-values to each action for every agent $u_{ij} \, | i = 1, 2, ..., n \quad j = 1, 2, ..., M$, the function's outputs are the elements of $\phi_{[i,d,j]}$. The fuzzy rule base proposed can be considered as a zero order Takagi-Sugeno rule base [34]:

$$\text{if } x \text{ is } \mu_d(x) \text{ then } q_{[i,1]} = \phi_{[i,d,1]}; ...; \; q_{[i,M]} = \phi_{[i,d,M]}$$

The approximate Q-value can be calculated by:

$$\tilde{Q}(x, \mathbf{u}) = \sum_{i=1}^{n} \sum_{d=1}^{N} \mu_d(x) \, \phi_{[i,d,j]} \tag{13}$$

The expression (13) is a linear parameterized approximation, the Q-values of a specified state-action couple is estimated through a weighted sum, where the weights are generated by the membership functions [35]. This approximator can be denoted by an approximator mapping

$$F = R^z \rightarrow \mathbf{Q} \tag{14}$$

where R^z is the parameter space, the parameter ϕ represents the approximation of the Q-function:

$$\tilde{Q}(x, \mathbf{u}) = [F(\phi)](x, \mathbf{u}) \tag{15}$$

Thus we do not need to store a great amount of Q-values for every pair $(x, \mathbf{u})$. Only z parameters in ϕ are needed. The mapping approximator F only represents a subset of $\mathbf{Q}$ [36].

From the point of view of reinforcement learning, a linear parameterized approximation of F are preferred since they make more suitable to analyze the theoretical aspect. This is the reason for our choosing of using a linear parameterized approximation ϕ, in this way, the normalized membership functions can be considered as basis functions [37].

The Expression (15) provides a $\tilde{Q}$ which is an approximate Q-function, in place of the exact Q-function Q, so the approximate $\tilde{Q}$ is supplied to the mapping H:

$$\bar{Q}_{k+1}(x, \mathbf{u}) = (H \circ F)(\phi_k)(x, \mathbf{u}) \tag{16}$$

Most of the time the Q-function $\bar{Q}$ is not able to be stored in a explicit way [38], as alternative, it has to be represented in an approximate form using a projection mapping $P : \mathbf{Q} \to R^z$

$$\phi_{k+1}(x, \mathbf{u}) = P(\bar{Q}_{k+1})(x, \mathbf{u}) \tag{17}$$

which makes certain that $\tilde{Q}(x, \mathbf{u}) = [F(\phi)](x, \mathbf{u})$ is as near as possible of $\bar{Q}(x, \mathbf{u})$ [39], in the sense of least square regression:

$$P(Q) = \phi^* \quad \phi^* = \arg\min_{\phi} \sum_{\lambda}^{s} (Q(x_\lambda, \mathbf{u}_\lambda) - [F(\phi)](x_\lambda, \mathbf{u}_\lambda))^2 \tag{18}$$

where a set of state-joint actions samples $(x, \mathbf{u})$ are used. Because of the use of triangular membership function shapes, and the linear parameterized approximation mapping F, (18) is a convex quadratic optimization problem where $z = nNM$ samples are used [40], so the expression (18) is reduced to a designation in the form:

$$\phi_{[i,d,j]} = P(Q)_{[i,d,j]} = Q(x, \mathbf{u}) \tag{19}$$

Recapitulating, the approximate linear fuzzy representation of the joint Q-function begins with an arbitrary value of the vector parameter vector ϕ and actualizes it in each iteration using the mapping:

$$\phi_{k+1} = (P \circ H \circ F)(\phi_k) \tag{20}$$

and stops when a parameter threshold ξ is greater than the difference between 2 consecutive parameters vector ϕ

$$\|\phi_{k+1} - \phi_k\| \leq \xi \tag{21}$$

A greedy policy can be obtained to control the system from ϕ^* (which is the parameter vector derived when $k \to \infty$), for whichever state, the actions are calculated by interpolation between the best local actions for each agent for every membership function center x_d:

$$h_i^*(x) = \sum_{d=1}^{N} \phi_{i,d}(x) \, u_{j_{id}^*} \quad where j_{i,d}^* \in \arg\max[F(\phi^*)](x, \mathbf{u}) \tag{22}$$

To implement the update (20), we propose a procedural using a modified version of the Q-learning algorithm [19], where the linear parameterization is added, in this way the algorithm can be extended to Multi-agent problems with continuous state space but with discrete action space. The algorithm starts with an arbitrary ϕ (it can be $\phi = 0$) until a threshold ξ is reached after several iterations.

2.5. Reinforcement Learning Algorithm for Continuous State Space

The linear fuzzy approximation depicted by (20) can be described by the following algorithm, where is used a modified version of Q-learning algorithm. To set a correspondence among the algorithm and the expression (20), the right hand of step 2 can be seen as (16) and then using the expression (17). Here the dynamics f, the reward function ρ and the discount factor γ are known in the form of a batch sample data.

1. Let $\alpha \in (0,1]$, $\epsilon \in (0,1]$ set

$$\phi(x, \mathbf{u}) \longleftarrow 0 \tag{23}$$

    where $(1 - \epsilon)$ is the probability of choose a greedy action in the state x, and ϵ is the probability of choose a random joint-action in $\mathbf{U}$.

2. Repeat in each iteration k

    - For state x, we select a joint action $\mathbf{U}$ with a suitable exploration. At each step a random action with probability $\varepsilon \in (0,1)$ is used.
    - Applying the linear fuzzy parameterization with membership functions μ_d $d = 1, ..., N$ and discrete actions U_j $j = 1, ..., M$, the threshold $\xi > 0$

$$\phi_{[i,d,j]_{k+1}} \longleftarrow \phi_{[i,d,j]_k} + \alpha_k \left[r_{k+1} + \beta - \Gamma \right] \Omega$$

$$\beta = \gamma \max_{j'} \sum_{i=1}^{n} \sum_{d'=1}^{N} \mu_{d'} \left(f \left(x_{k+1}, \mathbf{u'} \right) \right) \phi_{[i,d',j']} \tag{24}$$

$$\Gamma = \sum_{i=1}^{n} \sum_{d=1}^{N} \mu_d \left(f \left(x_k, \mathbf{u} \right) \right) \phi_{[i,d,j]} \tag{25}$$

$$\Omega = \sum_{i=1}^{n} \sum_{d=1}^{N} \mu_d \left(f \left(x_k, \mathbf{u} \right) \right) \tag{26}$$

    - Until:

$$\|\phi_{k+1} - \phi_k\| \leq \xi \tag{27}$$

    - Output:

$$\phi^* = \phi_{k+1} \tag{28}$$

A greedy policy is obtained to control the system by:

$$h_i^*(x) = \sum_{d=1}^{N} \phi_{i,d}(x) \, u_{j_{id}^*} \quad \text{where } j_{i,d}^* \in \arg\max[F(\phi^*)](x, \mathbf{u}) \tag{29}$$

where $j_{i,d}^* \in \arg\max F(\phi^*)(x, \mathbf{u})$, $j_{i,d}^*$ is the optimal action for the center x_d for the agent i.

Theorem 1. *The algorithm with linear fuzzy parameterization* (20) *converges to a fixed vector* ϕ^*.

Proof of Theorem 1. Since the mapping given by F,

$$[F(\phi)](x, \mathbf{u}) = \sum_{i=1}^{n} \sum_{d=1}^{N} \mu_d(x) \, \phi_{[i,d,j]} \tag{30}$$

the convergence of the algorithm is guaranteed through ensuring that the compound mapping $P \circ H \circ F$ is a contraction in the infinite norm. [28] shows that the mapping H is a contraction, so it remains to demonstrate that F and P are not expansions. The mapping given by F is a weighted linear combination of membership functions

$$\left\| [F(\phi)](x,\mathbf{u}) - [F(\phi')](x,\mathbf{u}) \right\|$$

$$= \left\| \sum_{i=1}^{n}\sum_{d=1}^{N} \mu_d(x)\,\phi_{[i,d,j]} - \sum_{i=1}^{n}\sum_{d=1}^{N} \mu_d(x)\,\phi'_{[i,d,j]} \right\|$$

$$= \left\| \sum_{i=1}^{n}\sum_{d=1}^{N} \mu_d(x)\left[\phi_{[i,d,j]} - \phi'_{[i,d,j]}\right] \right\|$$

$$= \left| \sum_{i=1}^{n}\sum_{d=1}^{N} \mu_d(x) \right| \left\| \phi_{[i,d,j]} - \phi'_{[i,d,j]} \right\|$$

$$\leq \sum_{i=1}^{n}\sum_{d=1}^{N} |\mu_d(x)| \left\| \phi_{[i,d,j]} - \phi'_{[i,d,j]} \right\|$$

$$\leq \sum_{i=1}^{n}\sum_{d=1}^{N} \mu_d(x) \left\| \phi_{[i,d,j]} - \phi'_{[i,d,j]} \right\|$$

$$\leq \sum_{i=1}^{n}\sum_{d=1}^{N} \mu_d(x) \left\| \phi - \phi' \right\|_{\infty}$$

$$\leq \left\| \phi - \phi' \right\|_{\infty}$$

$$(31)$$

where the last step is true because the sum of the standard functions $\mu_d(x)$ is 1, and the product generated by each agent also is 1. So it shows that the mapping F is a non-expansion. Since the mapping P is

$$P(Q)_{[i,d,j]} = Q(x,\mathbf{u}) \tag{32}$$

and the samples are centers of the membership functions $\phi_k(x_k,\mathbf{u}_k) = 1$, so the mapping P is a non-expansion [33]. Since H mapping is a contraction with $\gamma < 1$, so $P \circ H \circ F$ is also a contraction by the factor γ

$$\left\| (P \circ H \circ F)(\phi) - (P \circ H \circ F)(\phi') \right\| \leq \gamma \left\| \phi - \phi' \right\|_{\infty} \tag{33}$$

where $P \circ H \circ F$ has a fixed vector ϕ^*, and the algorithm above converges to this fixed point as $k \to \infty$. $\square$

Theorem 2. *For any choice of $\xi > 0$ and any initial threshold value parameter vector $\phi_0 \in R^z$, the algorithm with linear fuzzy parameterization is completed in a finite time.*

Proof of Theorem 2. As shown in Theorem 1, the mapping is a contraction $P \circ H \circ F$ with $\gamma < 1$ and a fixed vector ϕ^*

$$\begin{aligned} &\|\phi_{k+1} - \phi^*\|_{\infty} \\ &= \|(P \circ H \circ F)(\phi_k) - (P \circ H \circ F)(\phi^*)\| \\ &< \gamma \|\phi_k - \phi^*\|_{\infty} \end{aligned} \tag{34}$$

So, if $\|\phi_{k+1} - \phi^*\|_{\infty} < \gamma \|\phi_k - \phi^*\|_{\infty}$, for induction $\|\phi_k - \phi^*\|_{\infty} < \gamma^k \|\phi_0 - \phi^*\|$ for $k > 0$. According to Banach fixed point, ϕ^* is bounded. Since the vector where the iteration starts is bounded, then $\|\phi_0 - \phi^*\|_{\infty}$ is also bounded. Let $G_0 = \|\phi_0 - \phi^*\|_{\infty}$ which is bounded and $\|\phi_k - \phi^*\|_{\infty} \leq \gamma^k G_0$ for $k > 0$, applying the triangle inequality:

$$\begin{aligned} \|\phi_{k+1} - \phi_k\|_{\infty} &\leq \|\phi_{k+1} - \phi^*\|_{\infty} + \|\phi_k - \phi^*\|_{\infty} \\ &\leq \gamma^{k+1} G_0 + \gamma^k G_0 = \gamma^k G_0\,[\gamma + 1] \end{aligned} \tag{35}$$

If $\gamma^k G_0\,[\gamma + 1] = \xi$,

$$\gamma^k = \frac{\xi}{G_0\,[\gamma + 1]} \tag{36}$$

Applying γ log base on both side of the above expression

$$k = \log_\gamma \left[\frac{\xi}{G_0\,[\gamma + 1]} \right] \tag{37}$$

with $G_o = \|\phi_0 - \phi^*\|_\infty$ which is bounded and $\gamma < 1$ implies that k is finite. So the algorithm is arrived in the most k iterations. $\square$

3. Results

3.1. Simulation of a Cooperative Task with Mobile Robots

We perform a simulation where the linear fuzzy parameterization is applied to a two-dimensional Multi-agent cooperative problem with continuous states and discrete actions. The two agents with mass m have to be directed in a flat surface, such that they reach the origin at the same time with minimum time elapsed. The information available to the agents consists of the reward function, the transition function of states and joint actions.

In the simulation environment, the state $x = [x_1, x_2, ..., x_8]^T$ has the coordinates in two dimensions of each agent s_{ix}, s_{iy} and their velocities in two dimensions $\dot{s}_{ix}$, $\dot{s}_{iy}$ for $i = 1,2$: $x = [s_{1x}, s_{1y}, \dot{s}_{1x}, \dot{s}_{1y}, s_{2x}, s_{2y}, \dot{s}_{2x}, \dot{s}_{2y}]^T$. The continuous state space model of the simulated system is:

$$
\begin{aligned}
\ddot{s}_{1x} &= -\eta\,(s_{1x}, s_{1y}) \frac{\dot{s}_{1x}}{m_1} + \frac{u_{1x}}{m_1} \\
\ddot{s}_{1y} &= -\eta\,(s_{1x}, s_{1y}) \frac{\dot{s}_{1y}}{m_1} + \frac{u_{1y}}{m_1} \\
\ddot{s}_{2x} &= -\eta\,(s_{2x}, s_{2y}) \frac{\dot{s}_{2x}}{m_2} + \frac{u_{2x}}{m_2} \\
\ddot{s}_{2y} &= -\eta\,(s_{2x}, s_{2y}) \frac{\dot{s}_{2y}}{m_2} + \frac{u_{2y}}{m_2}
\end{aligned}
\tag{38}
$$

where $\eta\,(s_{ix}, s_{iy})$ for $i = 1,2$ is the function friction which depends of the position of each agent, the control signal is $\mathbf{U} = [u_{1x}, u_{1y}, u_{2x}, u_{2y}]^T$ which is a force and m_i for $i = 1,2$ is the mass of each robot.

The system is discretized with a step of $T = 0.4s$ and the expression that describes the dynamics of the system are integrated between the sampling time. In the task, we select the start points randomly and carry out 50 training iteration, in the case of reaching 50 iterations without accomplishing the final goal, the experiment is restarted.

The magnitude of the state and action variables are bounded. To make more tractable the problem, s_{ix} and $s_{iy} \in [-6,6]$ meters, $\dot{s}_{ix}$ and $\dot{s}_{iy} \in [-3,3]\,\frac{m}{s}$, also the force is bounded $u_{ix}, u_{ix} \in [-2,2]$ for $i = 1,2$, the friction coefficient is taken constant with $\eta = 1\,\frac{kg}{s}$, the mass of the agent is taken equal for both $m = 0.5$ kg.

The actions control for each agent are discrete with 25 elements $U_i = [-2\ -0.2\ 0\ 0.2\ 2] \times [-2\ -0.2\ 0\ 0.2\ 2]$ for $i = 1,2$, they correspond to force in diagonal, left, right, up, down and no force applied. The membership functions used for the position state and velocity state have triangular shape, where the core of each membership function is x_d . The cores of the membership function for the location domain s is centered at $[-6, -3, -0.3, -0.1, 0, 0.1, 0.3, 3, 6]$ and the cores of the membership function for the velocity domain are: $[-3, -1, 0, 1, 3]$, each one for every agent, this is shown in Figure 1. In this way 50625 pairs $(x, \mathbf{u})$ are storage for each agent in the vector parameter ϕ, this amount increases with the number of membership functions. An example of fuzzy triangular partition is showed in Figure 1.

The partition of the state space x is determined by the product of the individual membership function for each agent i

$$
\mu(x) = \prod_{i=1}^{2} \mu_{s_{ix}} \prod_{i=1}^{2} \mu_{s_{iy}} \prod_{i=1}^{2} \mu_{\dot{s}_{ix}} \prod_{i=1}^{2} \mu_{\dot{s}_{iy}}
\tag{39}
$$

The final objective of arriving at the same is shown by a common reward function ρ:

$$
\rho\,(x, \mathbf{u}) = 5 \text{ if } \|x\| < 0.1
\tag{40}
$$
$$
\rho\,(x, \mathbf{u}) = 0 \text{ in another way}
$$

As regards the coordination problem, the agents accomplish an implicit coordination, where they learn to prefer one solution about equally good solutions by chance and then overlook the other options [41].

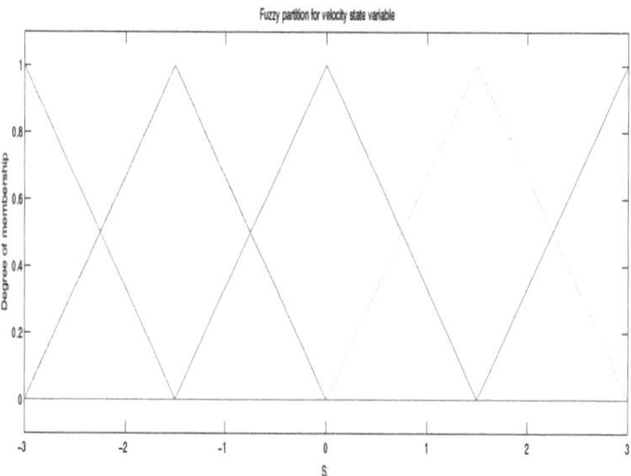

Figure 1. Triangular fuzzy partition for velocity state.

After the algorithm is performed and ϕ^* is obtained, a policy can be derived by interpolation between the best local action for each agent:

$$h_i^*(x) = \sum_{d=1}^{N} \phi_{i,d}(x)\, u_{j_{id}^*} \quad \text{where } j_{i,d}^* \in \arg\max F(\phi^*)(x, \mathbf{u}) \tag{41}$$

For the simulation, the learning parameters were set $\gamma = 0.96$ and the $\xi = 0.05$, the initial conditions for the experiment were set $s_0 = [-4, -6, -2, 2, 5, 3, 2, -1]$, the algorithm shows a convergence after 15 iterations, Figure 2 shows the states and the signal control $U_1 = [u_{1x}, u_{1y}]$ for the agent 1, Figure 3 shows the states and the signal control $U_2 = [u_{2x}, u_{2y}]$ for the agent 2. The final path followed by both agents are shown in Figure 4.

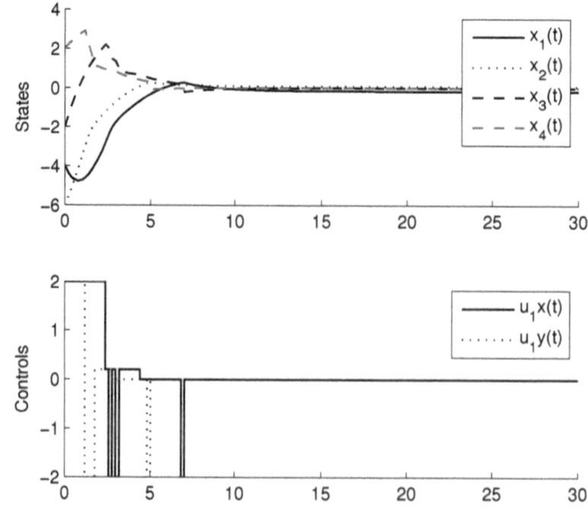

Figure 2. States and signal control for the Agent 1.

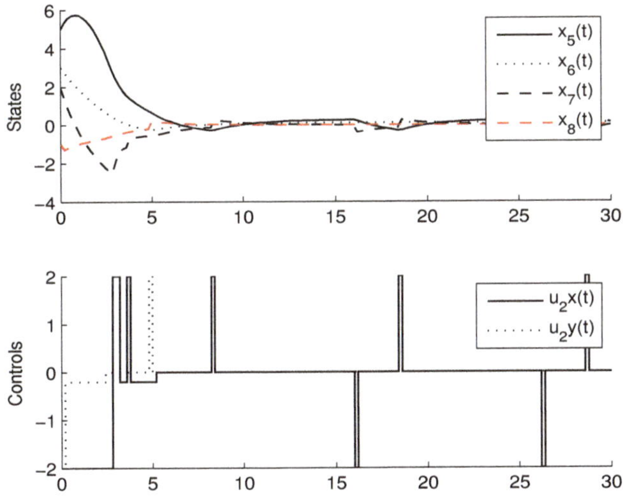

Figure 3. States and signal control for Agent 2.

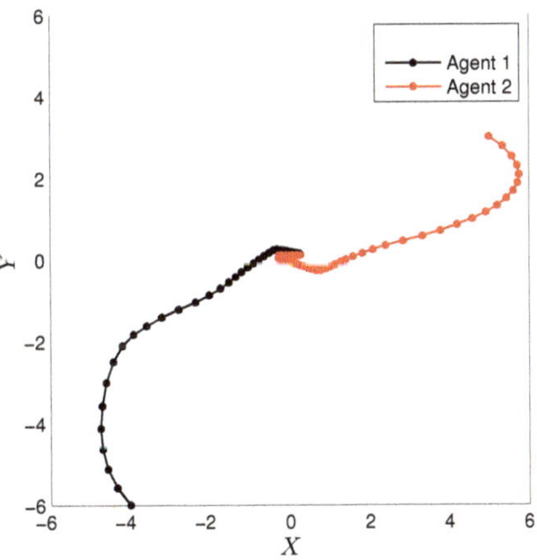

Figure 4. Final path by Agent 1 and Agent 2.

3.2. Experimental Set-up

Two mobile robots Khepera IV are used to perform a experiment in MAS [42]. They have 5 sensors which are placed around the robot and are positioned and numbered as shown in Figure 5. These 5 sensors are ultrasonic devices compose of one transmitter and one receiver, they are used to detect the physical features of the environment such as obstacles and other nearby agents.

The five Khepera's sonar readings $l_{a,c}$ are quantified in three degrees. They represent the amount of closeness to the nearest obstacle or others agents, 0 indicates obstacles or agents which are near, 1 indicates obstacles or agents which are in a medium distance and finally 2 indicates obstacles or agents which are relatively far from the sensors.

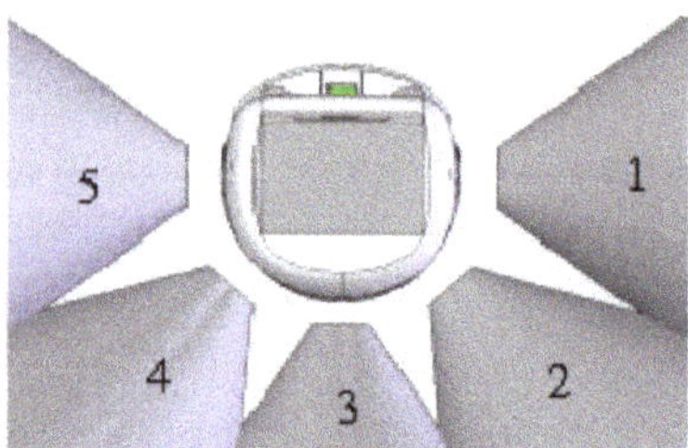

Figure 5. Position of the Khepera's UltraSonic sensors.

The parameters d_a (the distance to the target or the goal) and p_a (the relative angle to the target or goal) are divided in eight degrees (0–8). Where 0 represents the smallest distance or angle and 8 represents the greatest relative distance or angle from the current Khepera's position to the target or goal.

The actions available for the robot khepera are:

- Move forward
- Turn in clockwise direction
- Turn in counter clockwise direction
- Stand-Still

The ultra-sonic sensors on the Khepera are used to help the mobile robots to determine if there are any obstacles in the environment. The experimental set up reveals that the reinforcement learning algorithm relies strongly in the sensors readings.

Sensor readings in the ideal simulation situation are based on mathematical calculations which are accurate and consistent. In the experimental implementation the readings are inaccurate and fluctuating. During the application of the controller this effect is minimized by permitting a period after performing a joint action, with the above we ensure that the sensor has steady reading before it is recorded. In addition, by moving the robots at relatively slow step during the learning process, the collisions with other objects or agent are reduced. The quantified readings would be enough to represent the current location and velocity when the robots are moving [43].

3.3. Experimental Results

To validate the proposed algorithm, the linear fuzzy approximator of the joint Q-function is applied to a two-dimensional Multi-agent cooperative task. Two mobile robots Khepera IV must be driven in a surface such that both agents reach the origin at the same time with minimum time elapsed, it is shown in Figure 6.

The fuzzy partition and the location of centroids used for the states were the same as the simulation section.

The goal of arriving at the same moment toward the origin in minimum time elapsed is shown by the common reward function ρ:

$$\rho\left(x, \mathbf{u}\right) = 5 \text{ if } \|x\| < 0.2 \tag{42}$$

$$\rho\left(x, \mathbf{u}\right) = 0 \text{ in another way}$$

For this experiment the learning parameters were set $\gamma = 0.96$ and $\xi = 0.2$, the initial conditions for the experiment were set $s_0 = [-4, 5, 0, 0, 5, 3, 0, 0]$, the experiment shows a convergence after 27 iterations. Figure 7 shows the states, the signal control $U_1 = [u_{1x}, u_{1y}]$ and the rewards for the agent 1 and Figure 8 shows the states, the signal control $U_2 = [u_{2x}, u_{2y}]$ and the reward for the agent 2.

Figure 6. Experimental test

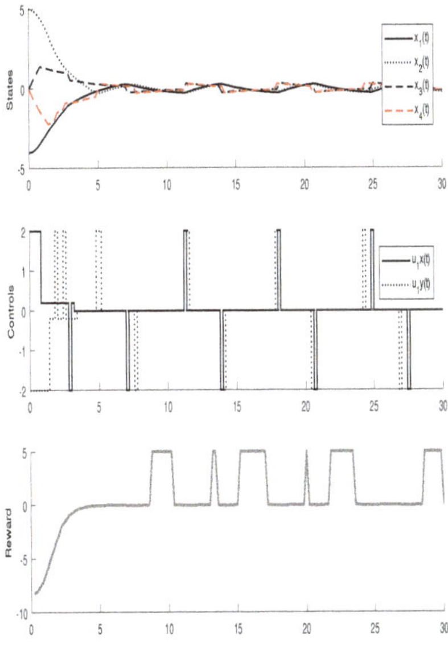

Figure 7. States and Rewards for the Agent 1 in the experimental implementation.

The election of the value for γ and ζ was set arbitrarily, the vector ϕ converged after 27 iterations when the bounded $\|\phi_{l+1} - \phi_l\| \leq \zeta$ was reached. The final path is shown in Figure 9, this trajectory is evidently different from the optimal policy, which would drive both agents in a straight line toward the goal since any initial position. However, with the fuzzy quantization used in this implementation and the effect of the damping, the final path obtained is the best that can be accomplished with this fuzzy parameterization. The coordination problem was overcame using an indirect method, where the agents learn to choose a solution bv chance.

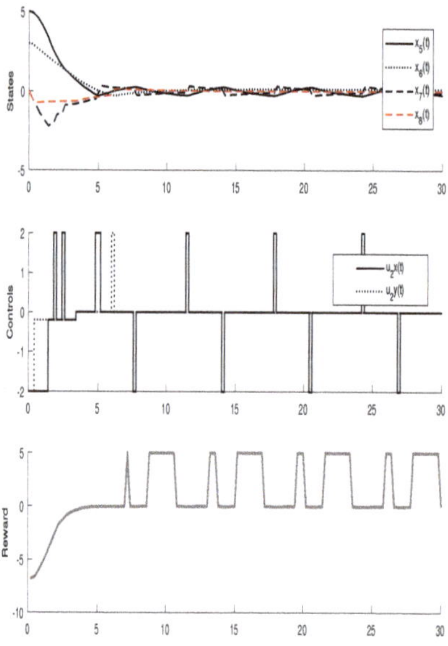

Figure 8. States and Rewards for the Agent 2 in the experimental implementation.

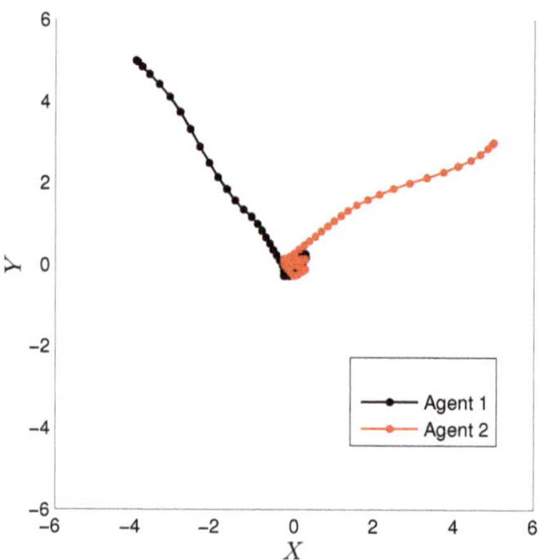

Figure 9. Final path by Agent 1 and Agent 2 in experimental implementation.

4. Comparison with CMOMMT Algorithm for Multi-Agent Systems

There are other methods for MARL in continuous state space, these proposals are restricted to a limited kind of task, one of this methods is "cooperative multi-robot observation of multiple moving target" (CMOMMT) given in [13], which relies in local information in order to learn cooperative behavior. It allows the application of the reinforcement learning in continuous state space for Multi-agents systems.

The kind of cooperation learned in this proposal is by implicit techniques, in this way this method is useful for reducing the representation of the state space through of a mapping of the continuous state space in a finite state space, where every new state discretized is considered as a region in the continuous state space.

The action space is discrete, this method uses a delayed reward where positive or negative values are obtained at the end of the training. Finally, an optimal joint policy of actions is obtained by a clustering technique in the discrete action space.

We performed the same cooperative task of arriving to the origin point at the same time for 2 agents as in the section above, using the same reward function and the same continuous state space. The initial condition was set $s_0 = [-4, 5, 0, 0, 5, -3, 0, 0]$.

The final path obtained by the CMOMMT algorithm is shown in Figure 10, where the path traced is not straight enough and near to the origin point is shown as a persistent oscillation. The signal state and the signal control for the agent 1 and agent 2 are shown in Figure 11 and the Figure 12, respectively.

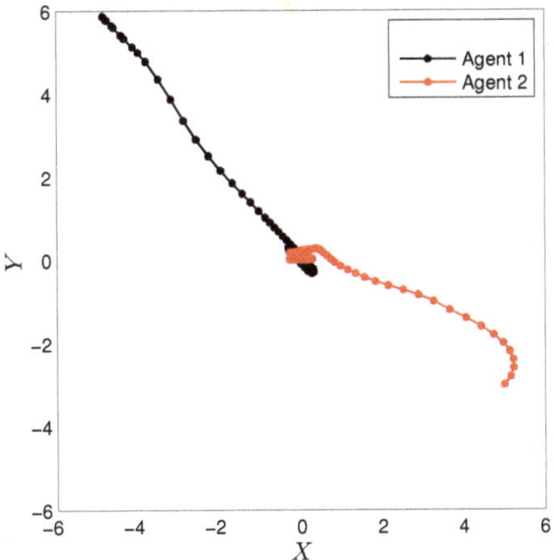

Figure 10. Final path obtained by CMOMMT algorithm.

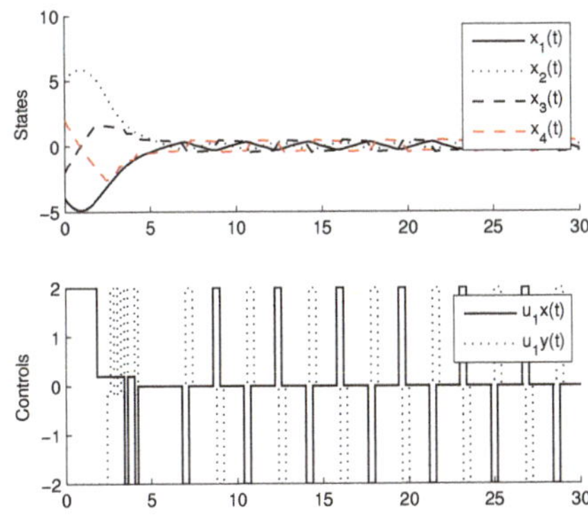

Figure 11. States and signal control for agent 1 CMOMMT algorithm.

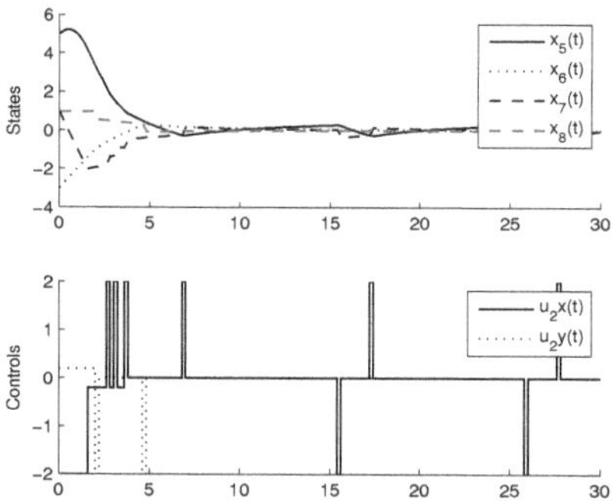

Figure 12. States and signal control for agent 2 CMOMMT algorithm.

The Table 1 shows a comparison between our proposal and the CMOMMT algorithm for an average of trials conducted under the same initial conditions.

Table 1. Comparison between fuzzy partition and CMOMMT.

Training	CMOMMT	Fuzzy Parameterization
Error (cm)	18	6
Time (s)	25	12
Iterations	34	27

One possible reason for the results of CMOMMT could be that the Q-functions obtained by this method are less smooth than the presented by the fuzzy parameterization. In this way the method proposed in our paper shows a better performance in the form of the more accuracy representation of the state space and less computational resources used by it.

5. Conclusions

One of the principal research direction of the artificial intelligent is to develop autonomous mobile robots with cooperative skills in continuous state space. For this reason, in this paper we have presented a linear fuzzy parameterization of the joint Q-function which is used with a modified version Q-learning algorithm. Our algorithm proposed can handle MARL problems with continuous state space, minimizing the time of convergence and avoiding storing the entire Q-values in a look up table. Triangular shapes were used to set the membership functions to define de estate space of the environment, this form was selected to simplify the projection mapping.

The main contribution of our work is that we present a reinforcement learning algorithm for MAS which uses a linear fuzzy parameterization of the joint Q-function. This approximation is carried out by means of a parameterization vector which only stores the Q-values at the centers of the triangular membership functions. The Q-values that are not found in the center are calculated by means of a weighted sum according to their degree of membership.

Two theorems were presented to guarantee the convergence to a fixed point in a finite number of iterations. The proposed method is a off-line model-based algorithm with deterministic dynamics, in the assumption that the joint reward function and the transition function are known by all the agents. Since having that kind of knowledge could be difficult in a real-life application, a future work could be to extend this proposal to a model free method, where the agents can learn by itself the dynamics of the environment, also this extension can be done to encompass problems with stochastic dynamics.

The performance of the linear fuzzy parameterization was evaluated first through simulation using Matlab software and then by an experiment where the task involves two mobile robots Khepera IV. Finally, the results obtained was compared with another suitable algorithm called CMOMMT which is capable of deal with tasks where the estate space is continuous.

Author Contributions: Formal analysis and Investigation D.L.-C.; Software, F.G.-L.; Validation, L.P.-D.; Review of the methods and editing, S.K.G.

Funding: The authors extend their appreciation to the Mexico's Secretary of Public Education for funding this work through research agreement 511-6/17-7605.

Conflicts of Interest: The authors declare no conflict of interest. The founding sponsors had no role in the design of the study; in the collection, analyses, or interpretation of data; in the writing of the manuscript, and in the decision to publish the results.

References

1. Sen, S.; Weiss, G. *Multiagent Systems: A Modern Approach to Distributed Artificial Intelligence*; MIT Press: Cambridge, MA, USA, 1999.
2. Stone, P.; Veloso, M. Multiagent systems: A survey from machine learning perspective. *Auton. Robots* **2000**, *8*, 345–383. [CrossRef]
3. Wooldridge, M. *An Introduction to MultiAgent Systems*; John Wiley & Sons: Hoboken, NL, USA, 2002.
4. Rashid, A.T.; Ali, A.A.; Frasca, M.; Fortuna, L. Path planning with obstacle avoidance based on visibility binary tree algorithm. *Robot. Auton. Syst.* **2013**, *61*, 1440–1449. [CrossRef]
5. Kaelbling, L.P.; Littman, M.L.; Moore, A.W. Reinforcement Learning: A Survey. *J. Artif. Intell. Res.* **1996**, *4*, 237–285. [CrossRef]
6. Arel, I.; Liu, C.; Urbanik, T.; Kohls, A.G. Reinforcement learning-based multi-agent system for network traffic signal control. *IET Intell. Transp. Syst.* **2010**, *4*, 128–135. [CrossRef]
7. Cherkassky, V.; Mulier, F. *Learning from data: Concepts, Theory and Methods*; Wiley-IEEE Press: Hoboken, NL, USA, 2007.
8. Barlow, S.V. Unsupervised learning. In *Unsupervised Learning: Foundations of Neural Computation*, 1st ed.; Sejnowski, T.J., Hinton, G., Eds.; MIT Press: Cambridge, MA, USA, 1999; pp. 1–18, ISBN 9780262581684.
9. Zhang, W.; Ma, L.; Li, X. Multi-agent reinforcement learning based on local communication. *Clust. Comput.* **2018**, 1–10. [CrossRef]
10. Hu, X.; Wang, Y. Consensus of Linear Multi-Agent Systems Subject to Actuator Saturation. *Int. J. Control Autom. Syst.* **2013**, *11*, 649–656. [CrossRef]
11. Luviano, D.; Yu, W. Path planning in unknown environment with kernel smoothing and reinforcement learning for multi-agent systems. In Proceedings of the 12th International Conference on Electrical Engineering, Computing Science and Automatic Control (CCE), Mexico City, Mexico, 28–30 October 2015.
12. Abul, O.; Polat, F.; Alhajj, R. Multi-agent reinforcement learning using function approximation. *IEEE Trans. Syst. Man Cybern. Part C Appl. Rev.* **2000**, 485–497. [CrossRef]
13. Fernandez, F.; Parker, L.E. Learning in large cooperative multi-robots systems. *Int. J. Robot. Autom. Spec. Issue Comput. Intell. Tech. Coop. Robots* **2001**, *16*, 217–226.
14. Foerster, J.; Nardelli, N.; Farquhar, G.; Afouras, T.; Torr, P.H.; Kohli, P.; Whiteson, S. Stabilising experience replay for deep multi-agent reinforcement learning. *arXiv* **2017**, arXiv:1702.08887.
15. Tamakoshi, H.; Ishi, S. Multi agent reinforcement learning applied to a chase problem in a continuous world. *Artif. Life Robot.* **2001**, 202–206. [CrossRef]
16. Ishiwaka, Y.; Sato, T.; Kakazu, Y. An approach to pursuit problem on a heterogeneous multiagent system using reinforcement learning. *Robot. Auton. Syst.* **2003**, *43*, 245–256. [CrossRef]
17. Radac, M.-B.; Precup, R.-E.; Roman, R.-C. Data-driven model reference control of MIMO vertical tank systems with model-free VRFT and Q-Learning. *ISA Trans.* **2017**. [CrossRef] [PubMed]
18. Pandian, B.J.; Noel, M.M. Control of a bioreactor using a new partially supervised reinforcement learning algorithm. *J. Process Control* **2018**, *69*, 16–29. [CrossRef]
19. Watkins, C.J.; Dayan, P. Q-learning. *Mach. Learn.* **1992**, *8*, 279–292. [CrossRef]
20. Nguyen, T.; Nguyen, N.D.; Nahavandi, S. Multi-Agent Deep Reinforcement Learning with Human Strategies. *arXiv* **2018**, arXiv:1806.04562.

21. Boutilier, C. Planning, Learning and Coordination in Multiagent Decision Processes. In Proceedings of the Sixth Conference on Theoretical Aspects of Rationality and Knowledge (TARK96), De Zeeuwse Stromen, The Netherlands, 17–20 March 1996; pp. 195–210.

22. Harsanyi, J.C.; Selten, R. *A General Theory of Equilibrium Selection in Games*, 1st ed.; MIT Press: Cambridge, MA, USA, 1988; ISBN 9780262081733.

23. Busoniu, L.; De Schutter, B.; Babuska, R. Decentralized Reinforcement Learning Control of a robotic Manipulator. In Proceedings of the International Conference on Control, Automation, Robotics and Vision, Singapore, 5–8 December 2006.

24. Littman, M.L. Value-function reinforcement learning in Markov games. *J. Cogn. Syst. Res.* **2001**, *2*, 55–66. [CrossRef]

25. Guestrin, C.; Lagoudakis, M.G.; Parr, R. Coordinated reinforcement learning. In Proceedings of the 19th International Conference on Machine Learning (ICML-2002), Sydney, Australia, 8–12 July 2002; pp. 227–234.

26. Bowling, M.; Veloso, M. Multiagent learning using a variable learning rate. *Artif. Intell.* **2002**, *136*, 215–250. [CrossRef]

27. Bertsekas, D.P. *Dynamic Programming and optimal control vol. 2*, 4th ed.; Athena Scientific: Belmont, MA, USA, 2017; ISBN 1-886529-44-2.

28. Istratesku, V. *Fixed Point Theory: An introduction*; Springer: Berlin, Germany, 2002; ISBN 978-1-4020-0301-1.

29. Melo, F.S.; Meyn, S.P.; Ribeiro, M.I. An analysis of reinforcement learning with functions approximation. In Proceedings of the 25th International Conference on Machine Learning (ICML-08), Helsinki, Finland, 5–9 July 2008; pp. 664–671.

30. Szepesvari, C.; Smart, W.D. Interpolation-based Q-learning. In Proceedings of the 21st International Conference on Machine Learning (ICML-04), Banff, AB, Canada, 4–8 July 2004; pp. 791–798.

31. Sutton, R.S.; McAllester, D.A.; Singh, S.P.; Mansour, Y. Policy gradient methods for reinforcement learning with function approximation. In Proceedings of the 12th International Conference on Neural Information Processing Systems, Denver, CO, USA, 29 November–4 December 1999; pp. 1057–1063.

32. Bertsekas, D.P.; Tsitsiklis, J.N. *Neuro-Dynamic Programming*, 1st ed.; Athena Scientific: Belmont, MA, USA, 1996; ISBN 1-886529-10-8.

33. Tsitsiklis, J.N.; Van Roy, B. Feature-based methods for large scale dynamic programming. *Mach. Learn.* **1996**, *22*, 59–94. [CrossRef]

34. Kruse, R.; Gebhardt, J.E.; Klowon, F. *Foundations of Fuzzy Systems*, 1st ed.; John Wiley & Sons: Hoboken, NL, USA, 1994; ISBN 047194243X.

35. Gordon, G.J. Reinforcement learning with function approximation converges to a region. *Adv. Neural Inf. Process. Syst.* **2001**, *13*, 1040–1046.

36. Tsitsiklis, J.N. Asynchronous stochastic approximation and Q-learning. *Mach. Learn.* **1994**, *16*, 185–202. [CrossRef]

37. Berenji, H.R.; Khedkar, P. Learning and tuning fuzzy logic controllers through reinforcements. *IEEE Trans. Neural Netw.* **1992**, *3*, 724–740. [CrossRef] [PubMed]

38. Munos, R.; Moore, A. Variable-resolution discretization in optimal control. *Mach. Learn.* **2002**, *49*, 291–323. [CrossRef]

39. Chow, C.S.; Tsitsiklis, J.N. An optimal one-way multi grid algorithm for discrete-time stochastic control. *IEEE Trans. Autom. Control* **1991**, *36*, 898–914. [CrossRef]

40. Busoniu, L.; Ernst, D.; De Schutter, B.; Babuska, R. Approximate Dynamic programming with fuzzy parametrization. *Automatica* **2010**, *46*, 804–814. [CrossRef]

41. Vlassis, N. A concise Introduction to Multi Agent Systems and Distributed Artificial Intelligence. In *Synthesis Lectures in Artificial Intelligence and Machine Learning*; Morgan & Claypool Publishers: San Rafael, CA, USA, 2007.

42. K-Team Corporation. 2013. Available online: http://www-k-team.com (accessed on 15 January 2018).

43. Ganapathy, V.; Soh, C.Y.; Lui, W.L.D. Utilization of webots and Khepera II as a Platform for neural Q-learning controllers. In Proceedings of the IEEE Symposium on Industrial Electronics and Applications, Kuala Lumpur, Malaysia, 25–27 May 2009.

Article

Time-Fractional Heat Conduction in a Plane with Two External Half-Infinite Line Slits under Heat Flux Loading

Yuriy Povstenko [1,*] **and Tamara Kyrylych** [2]

1   Faculty of Science and Technology, Jan Dlugosz University in Czestochowa, Armii Krajowej 13/15, 42-200 Czestochowa, Poland
2   Department of Law, Administration and Management, Jan Dlugosz University in Czestochowa, Zbierskiego 2/4, 42-200 Czestochowa, Poland; tamara.kyrylych@gmail.com
*   Correspondence: j.povstenko@ujd.edu.pl; Tel.: +48-343-612-269

Received: 22 April 2019; Accepted: 15 May 2019; Published: 18 May 2019

Abstract: The time-fractional heat conduction equation follows from the law of conservation of energy and the corresponding time-nonlocal extension of the Fourier law with the "long-tail" power kernel. The time-fractional heat conduction equation with the Caputo derivative is solved for an infinite plane with two external half-infinite slits with the prescribed heat flux across their surfaces. The integral transform technique is used. The solution is obtained in the form of integrals with integrand being the Mittag–Leffler function. A graphical representation of numerical results is given.

Keywords: fractional calculus; Caputo derivative; generalized Fourier law; Laplace transform; Fourier transform; Mittag–Leffler function

---

1. Introduction

The conventional Fourier law states the linear dependence between the heat flux vector $\mathbf{q}$ and the gradient of temperature:

$$\mathbf{q} = -k \operatorname{grad} T, \tag{1}$$

with k being the thermal conductivity of a solid.

The Fourier law (1) in combination with the law of conservation of energy

$$C\frac{\partial T}{\partial t} = -\operatorname{div} \mathbf{q} \tag{2}$$

leads to the classical parabolic heat conduction equation

$$\frac{\partial T}{\partial t} = a\Delta T, \tag{3}$$

where $a = k/C$ denotes the thermal diffusivity coefficient, C is the heat capacity,

In non-classical theories, the Fourier law (1) and the heat conduction Equation (3) are substituted by alternative equations which give an account of the complex internal structure of a body and reflect processes taking place at the microscopic level. Gurtin and Pipkin [1] generalized the Fourier law to time-nonlocal dependence between the heat flux vector $\mathbf{q}$ and the gradient of temperature. Nigmatullin [2] proposed the time-nonlocal extension of the Fourier law in the form:

$$\mathbf{q}(t) = -k \int_0^t K(t - \tau) \operatorname{grad} T(\tau) \, \mathrm{d}\tau, \tag{4}$$

with $K(t - \tau)$ being the time-nonlocality kernel. With the law of conservation of energy (2), the constitutive Equation (4) gives the heat conduction equation with memory [2]

$$\frac{\partial T}{\partial t} = a \int_0^t K(t - \tau) \, \Delta T(\tau) \, d\tau. \tag{5}$$

It is obvious that the standard Fourier law (1) and the parabolic heat conduction Equation (3) are obtained when

$$K(t - \tau) = \delta(t - \tau), \tag{6}$$

where $\delta(t - \tau)$ is the Dirac delta function.

Different extended theories of heat conduction correspond to the appropriate selection of the memory kernel $K(t - \tau)$. "Full sclerosis" (localized heat conduction) [3,4] is described by the memory kernel being the time derivative of the Dirac delta function

$$K(t - \tau) = -\delta'(t - \tau) \tag{7}$$

and results in the Helmholtz equation for temperature

$$T = a\Delta T. \tag{8}$$

"Full memory" [5,6] infers that there is no fading of memory, the kernel $K(t - \tau)$ is constant

$$K(t - \tau) = 1, \tag{9}$$

and temperature obeys the wave equation

$$\frac{\partial^2 T}{\partial t^2} = a\Delta T. \tag{10}$$

"Short-tail" memory [7,8] is associated with the exponential time-nonlocality kernel and leads to the Cattaneo telegraph equation for temperature. "Middle-tail" memory [9,10] involves kernels being different Mittag–Leffler functions and leads to the time-fractional telegraph equations.

The time-nonlocal dependences between the heat flux vector $\mathbf{q}$ and the gradient of temperature with the "long-tail" power kernel [3,4,11–14]

$$\mathbf{q}(t) = -\frac{k}{\Gamma(\alpha)} \frac{d}{dt} \int_0^t (t - \tau)^{\alpha-1} \operatorname{grad} T(\tau) \, d\tau, \qquad 0 < \alpha \le 1, \tag{11}$$

$$\mathbf{q}(t) = -\frac{k}{\Gamma(\alpha - 1)} \int_0^t (t - \tau)^{\alpha-2} \operatorname{grad} T(\tau) \, d\tau, \qquad 1 < \alpha \le 2, \tag{12}$$

can be interpreted in terms of fractional calculus:

$$\mathbf{q}(t) = -kD_{RL}^{1-\alpha} \operatorname{grad} T(t), \qquad 0 < \alpha \le 1, \tag{13}$$

$$\mathbf{q}(t) = -kI^{\alpha-1} \operatorname{grad} T(t), \qquad 1 < \alpha \le 2, \tag{14}$$

and in combination with the law of conservation of energy (2) result in the time-fractional heat conduction equation

$$\frac{\partial^\alpha T}{\partial t^\alpha} = a \, \Delta T, \qquad 0 < \alpha \le 2, \tag{15}$$

with the Caputo fractional derivative.

Here, $I^\alpha f(t)$ and $D^\alpha_{RL} f(t)$ are the Riemann–Liouville fractional integral and the Riemann–Liouville fractional derivative [15,16]

$$I^\alpha f(t) = \frac{1}{\Gamma(\alpha)} \int_0^t (t-\tau)^{\alpha-1} f(\tau)\, d\tau, \qquad \alpha > 0, \tag{16}$$

$$D^\alpha_{RL} f(t) = \frac{d^n}{dt^n}\left[\frac{1}{\Gamma(n-\alpha)} \int_0^t (t-\tau)^{n-\alpha-1} f(\tau)\, d\tau \right], \qquad n-1 < \alpha < n. \tag{17}$$

$\dfrac{d^\alpha f(t)}{dt^\alpha}$ is the Caputo fractional derivative

$$\frac{d^\alpha f(t)}{dt^\alpha} = \frac{1}{\Gamma(n-\alpha)} \int_0^t (t-\tau)^{n-\alpha-1} \frac{d^n f(\tau)}{d\tau^n}\, d\tau, \qquad n-1 < \alpha < n. \tag{18}$$

$\Gamma(\alpha)$ denotes the Gamma function.

Equation (15) combines the whole spectrum of equations starting with the localized heat conduction (the elliptic Helmholtz equation for temperature when $\alpha \to 0$) through the standard parabolic heat conduction equation (when $\alpha = 1$) and ending with the ballistic heat conduction (the hyperbolic wave equation for temperature when $\alpha = 2$).

Fractional calculus (the theory of integrals and derivatives of non-integer order) has many important applications in physics, geophysics, chemistry, continuum mechanics, rheology, engineering, biology, etc. (see, for example, Refs. [17–27] and references therein).

Starting from the fundamental papers by Wyss [28] and Mainardi [29,30], considerable interest has been shown in solving the time-fractional heat conduction equation (diffusion-wave equation) (15). The book [31] sums up investigations in this field.

Strength of solids depends significantly on the presence in a real solid such defects as cracks, slits, holes, inclusions, etc. During deformation in the vicinity of such defects, there appears considerable stress concentration which can lead to formation of new and growth of existing crack and, hence, to local or global fracture of a solid. Thermal shock conditions which can arise during sudden cooling of a solid can result in very high thermal stresses near a crack, therefore, thermal shock is an important fracture mechanism of brittle materials. The pioneering papers [32,33] laid the groundwork for study of thermoelastic problems for solids with cracks and slits based on the classical heat conduction Equation (3). The literature on this subject is quite voluminous (see, for example, Refs. [34–42] and references therein). More general heat conduction equations imply formulation of associated theories of thermal stresses. Thermoelasticity associated with the time-fractional heat conduction Equation (15) was proposed in [11]; the book [26] sums up investigations in this field. The first paper on fractional heat conduction in a solid with a crack was published by the authors [43]. In the present paper, the time-fractional heat conduction equation with the Caputo derivative is solved for an infinite plate with two external half-infinite slits whose surfaces are exposed to the heat flux loading. The thermal stress aspect of the problem including investigation of the stress intensity factor will be studied in future research.

2. Statement of the Problem

Consider a plane with two slits located on the half-lines $y = 0$, $-\infty < x < -l$ and $y = 0$, $l < x < \infty$. To study the heat conduction problem, we examine the time-fractional heat conduction equation

$$\frac{\partial^\alpha T}{\partial t^\alpha} = a \left(\frac{\partial^2 T}{\partial x^2} + \frac{\partial^2 T}{\partial y^2} \right),$$

$$-\infty < x < \infty, \quad -\infty < y < \infty, \quad 0 < t < \infty, \quad 0 < \alpha \le 2, \tag{19}$$

and assume zero initial conditions

$$t = 0: \quad T = 0, \qquad 0 < \alpha \leq 2, \tag{20}$$

$$t = 0: \quad \frac{\partial T}{\partial t} = 0, \quad 1 < \alpha \leq 2, \tag{21}$$

and heat flux loading according to the constitutive Equations (13) and (14)

$$y = 0^+: \quad kD_{RL}^{1-\alpha}\frac{\partial T}{\partial y} = q^+(x,t), \qquad l < |x| < \infty, \qquad 0 < \alpha \leq 1, \tag{22}$$

$$y = 0^+: \quad kI^{\alpha-1}\frac{\partial T}{\partial y} = q^+(x,t), \qquad l < |x| < \infty, \qquad 1 < \alpha \leq 2, \tag{23}$$

$$y = 0^-: \quad -kD_{RL}^{1-\alpha}\frac{\partial T}{\partial y} = q^-(x,t), \qquad l < |x| < \infty, \qquad 0 < \alpha \leq 1, \tag{24}$$

$$y = 0^-: \quad -kI^{\alpha-1}\frac{\partial T}{\partial y} = q^-(x,t), \qquad l < |x| < \infty, \qquad 1 < \alpha \leq 2. \tag{25}$$

The following conditions at infinity are imposed

$$\lim_{y \to \pm\infty} T(x,y,t) = 0, \tag{26}$$

$$\lim_{t \to \infty} T(x,y,t) = 0. \tag{27}$$

In the subsequent text, we will restrict our consideration to the problem with constant heat flux loading

$$q^+(x,t) = q^-(x,t) = q_0 = \text{const.} \tag{28}$$

In this case, due to the symmetry of the problem with respect to the x-axis, we can study only the upper half-plane $y \geq 0$. The problem is reformulated as follows: solve the heat conduction equation

$$\frac{\partial^\alpha T}{\partial t^\alpha} = a\left(\frac{\partial^2 T}{\partial x^2} + \frac{\partial^2 T}{\partial y^2}\right),$$

$$-\infty < x < \infty, \quad 0 < y < \infty, \quad 0 < t < \infty, \quad 0 < \alpha \leq 2, \tag{29}$$

with zero initial conditions

$$t = 0: \quad T = 0, \qquad 0 < \alpha \leq 2, \tag{30}$$

$$t = 0: \quad \frac{\partial T}{\partial t} = 0, \quad 1 < \alpha \leq 2, \tag{31}$$

under constant heat flux loading

$$y = 0: \quad kD_{RL}^{1-\alpha}\frac{\partial T}{\partial y} = q_0, \qquad l < |x| < \infty, \qquad 0 < \alpha \leq 1, \tag{32}$$

$$y = 0: \quad kI^{\alpha-1}\frac{\partial T}{\partial y} = q_0, \qquad l < |x| < \infty, \qquad 1 < \alpha \leq 2, \tag{33}$$

$$y = 0: \quad \frac{\partial T}{\partial y} = 0, \qquad 0 \leq |x| < l, \qquad 0 < \alpha \leq 2, \tag{34}$$

and zero conditions at infinity

$$\lim_{y\to\infty} T(x,y,t) = 0, \tag{35}$$

$$\lim_{t\to\infty} T(x,y,t) = 0. \tag{36}$$

3. Solution of the Problem

To solve the problem, we will employ the Laplace transform with respect to time t, the exponential Fourier transform with respect to the spatial coordinate x, and the cos-Fourier transform with respect to the spatial coordinate y.

Recall the Laplace transform rules for fractional integrals and derivatives [15,16]:

$$\mathcal{L}\{I^\alpha f(t)\} = \frac{1}{s^\alpha}\, f^*(s), \tag{37}$$

$$\mathcal{L}\{D^\alpha_{RL} f(t)\} = s^\alpha f^*(s) - \sum_{k=0}^{n-1} D^k I^{n-\alpha} f(0^+) s^{n-1-k}, \qquad n-1 < \alpha < n, \tag{38}$$

$$\mathcal{L}\left\{\frac{\mathrm{d}^\alpha f(t)}{\mathrm{d}t^\alpha}\right\} = s^\alpha f^*(s) - \sum_{k=0}^{n-1} f^{(k)}(0^+) s^{\alpha-1-k}, \qquad n-1 < \alpha < n. \tag{39}$$

Here, s denotes the Laplace transform variable; the asterisk marks the transform.

The integral transform technique results in the solution in the transform domain

$$\tilde{\tilde{T}}^*(\xi,\eta,s) = -\frac{\sqrt{2}aq_0}{k\sqrt{\pi}}\left[\pi\delta(\xi) - \frac{\sin(l\xi)}{\xi}\right]\frac{s^{\alpha-2}}{s^\alpha + a\left(\xi^2 + \eta^2\right)}. \tag{40}$$

Here, both Fourier transforms are marked by the tilde, ξ is the transform variable of the exponential Fourier transform, η is the transform variable of the cos-Fourier transform, $\delta(\xi)$ is the Dirac delta function appearing in the integral

$$\int_{-\infty}^{\infty} \cos(x\xi)\,\mathrm{d}x = 2\pi\delta(\xi). \tag{41}$$

Inverting the integral transforms with taking into account the formula for the inverse Laplace transform [15,16]

$$\mathcal{L}^{-1}\left\{\frac{s^{\alpha-\beta}}{s^\alpha + b}\right\} = t^{\beta-1} E_{\alpha,\beta}(-bt^\alpha), \tag{42}$$

where $E_{\alpha,\beta}(z)$ is the Mittag–Leffler function in two parameters α and β

$$E_{\alpha,\beta}(z) = \sum_{k=0}^{\infty} \frac{z^k}{\Gamma(\alpha k + \beta)}, \qquad \alpha > 0, \quad \beta > 0, \quad z \in C, \tag{43}$$

we get the solution

$$T(x,y,t) = -\frac{2aq_0 t}{\pi k}\int_0^\infty E_{\alpha,2}\left(-a\eta^2 t^\alpha\right)\cos(y\eta)\,\mathrm{d}\eta$$

$$+ \frac{2aq_0 t}{\pi^2 k}\int_0^\infty\int_{-\infty}^\infty E_{\alpha,2}\left[-a\left(\xi^2+\eta^2\right)t^\alpha\right]\frac{\sin(l\xi)}{\xi}\cos(x\xi)\cos(y\eta)\,\mathrm{d}\xi\,\mathrm{d}\eta. \tag{44}$$

To simplify the second term in Equation (44), we will use the polar coordinate system in the (η,ξ)-plane:

$$\xi = \rho\cos\theta, \qquad \eta = \rho\sin\theta \tag{45}$$

and the substitution $u = \cos\theta$. Having regard to the following integral [31]

$$\int_0^1 \frac{\sin(qw)\,\cos(p\sqrt{1-w^2})}{w\sqrt{1-w^2}}\,\mathrm{d}w = \frac{\pi}{2}\int_0^q J_0\left(\sqrt{p^2+z^2}\right)\mathrm{d}z, \tag{46}$$

where $J_n(z)$ is the Bessel function of the first kind; after standard mathematical treatment, we finally obtain

$$\begin{aligned}
T(x,y,t) &= -\frac{2aq_0t}{\pi k}\int_0^\infty E_{\alpha,2}\left(-a\eta^2 t^\alpha\right)\cos(y\eta)\,\mathrm{d}\eta \\
&\quad + \frac{aq_0t}{\pi k}\int_0^\infty \rho E_{\alpha,2}\left(-a\rho^2 t^\alpha\right) \\
&\quad \times \int_0^1 \left\{(l+x)J_0\left[\rho\sqrt{y^2+(l+x)^2u^2}\right] + (l-x)J_0\left[\rho\sqrt{y^2+(l-x)^2u^2}\right]\right\}\mathrm{d}u\,\mathrm{d}\rho. \tag{47}
\end{aligned}$$

Now, we present several special cases of the solution (48) associated with integer values of α.

3.1. Localized Heat Conduction ($\alpha \to 0$)

In this case,

$$E_{0,2}(-x) = \frac{1}{1+x}. \tag{48}$$

Taking into account that [44,45]

$$\int_0^\infty \frac{1}{x^2+c^2}\cos(bx)\,\mathrm{d}x = \frac{\pi}{2c}\,e^{-bc}, \quad b>0,\ c>0, \tag{49}$$

$$\int_0^\infty \frac{x}{x^2+c^2}J_0(bx)\,\mathrm{d}x = K_0(bc), \quad b>0,\ c>0, \tag{50}$$

where $K_n(z)$ is the modified Bessel function, we get

$$\begin{aligned}
T(x,y,t) &= -\frac{q_0\sqrt{at}}{k}\exp\left(-\frac{y}{\sqrt{a}}\right) + \frac{q_0t}{k\pi}\int_0^1\left\{(l+x)K_0\left[\sqrt{\frac{y^2+(l+x)^2u^2}{a}}\right]\right. \\
&\quad \left. + (l-x)K_0\left[\sqrt{\frac{y^2+(l-x)^2u^2}{a}}\right]\right\}\mathrm{d}u. \tag{51}
\end{aligned}$$

3.2. Classical Heat Conduction ($\alpha = 1$)

For the standard heat conduction equation,

$$E_{1,2}(-x) = \frac{1-e^{-x}}{x}. \tag{52}$$

Taking into consideration the following integrals evaluated by the authors,

$$\int_0^\infty \frac{1-e^{-cx^2}}{x^2}\cos(bx)\,\mathrm{d}x = \sqrt{\pi c}\left[\exp\left(-\frac{b^2}{4c}\right) - \frac{\sqrt{\pi}b}{2\sqrt{c}}\,\mathrm{erfc}\left(\frac{b}{2\sqrt{c}}\right)\right], \quad b>0,\ c>0, \tag{53}$$

$$\int_0^\infty \frac{1-e^{-cx^2}}{x}J_0(bx)\,\mathrm{d}x = \frac{1}{2}\,E_1\left(\frac{b^2}{4c}\right), \quad b>0,\ c>0, \tag{54}$$

we arrive at

$$T(x,y,t) = -\frac{2q_0\sqrt{at}}{k}\left[\frac{1}{\sqrt{\pi}}\exp\left(-\frac{y^2}{4at}\right) - \frac{y}{2\sqrt{at}}\,\text{erfc}\left(\frac{y}{2\sqrt{at}}\right)\right]$$

$$+ \frac{q_0}{2\pi k}\int_0^1\left\{(l+x)\mathbf{E}_1\left[\frac{y^2+(l+x)^2u^2}{4at}\right] + (l-x)\mathbf{E}_1\left[\frac{y^2+(l-x)^2u^2}{4at}\right]\right\}du. \tag{55}$$

Here, erfc (z) is the complementary error function, and $\mathbf{E}_n(z)$ is the exponential integral [46]

$$\mathbf{E}_n(z) = \int_1^\infty \frac{e^{-zt}}{t^n}\,dt. \tag{56}$$

We have changed the standard notation of the exponential integral $\mathbf{E}_n(z)$ to prevent confusion with the Mittag–Leffler function $E_\alpha(z)$.

3.3. Ballistic Heat Conduction ($\alpha = 2$)

In the case of the wave equation for temperature,

$$E_{2,2}(-x) = \frac{\sin\sqrt{x}}{\sqrt{x}}, \tag{57}$$

the required integrals read

$$\int_0^\infty \frac{\sin(cx)\cos(bx)}{x}\,dx = \begin{cases} \pi/2, & 0 < b < c, \\ 0, & 0 < c < b, \end{cases} \tag{58}$$

$$\int_0^\infty \sin(cx)\,J_0(bx)\,dx = \begin{cases} \dfrac{1}{\sqrt{c^2-b^2}}, & 0 < b < c, \\ 0, & 0 < c < b, \end{cases} \tag{59}$$

and using the Heaviside step function the solution takes the form

$$T(x,y,t) = -\frac{\sqrt{a}q_0}{k}H\left(\sqrt{a}t - y\right)$$

$$+ \frac{\sqrt{a}q_0}{\pi k}\int_{x-l}^{x+l}\left\{\begin{array}{ll}\dfrac{1}{\sqrt{at^2-y^2-u^2}}, & u^2 < at^2 - y^2 \\ 0, & u^2 > al^2 - y^2\end{array}\right\}du. \tag{60}$$

The dependence of the solution on the spatial coordinate x is shown in Figures 1 and 2; the dependence of the solution on the spatial coordinate y is depicted in Figures 3 and 4. In calculations, we have used the non-dimensional quantities

$$\bar{x} = \frac{x}{l}, \qquad \bar{y} = \frac{y}{l}, \qquad \kappa = \frac{\sqrt{a}t^{\alpha/2}}{l}, \qquad \bar{T} = \frac{kl}{aq_0t}T. \tag{61}$$

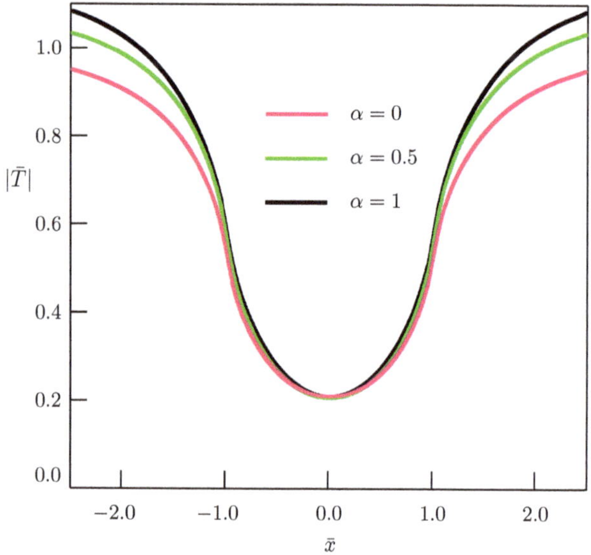

Figure 1. Dependence of the solution on the spatial coordinate x. Typical curves for $0 \leq \alpha \leq 1$; $\kappa = 1$.

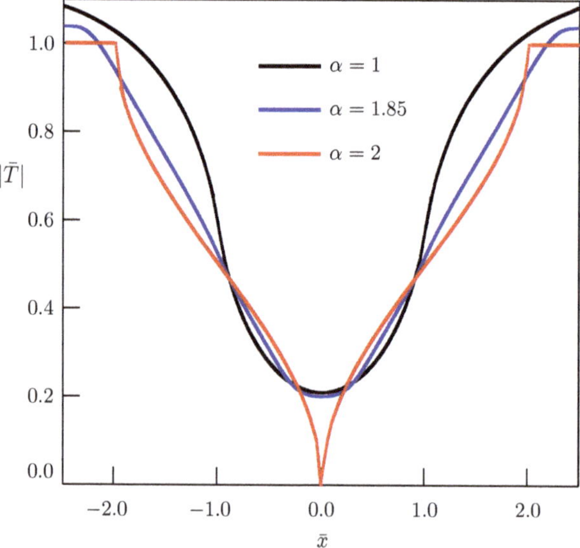

Figure 2. Dependence of the solution on the spatial coordinate x. Typical curves for $1 \leq \alpha \leq 2$; $\kappa = 1$.

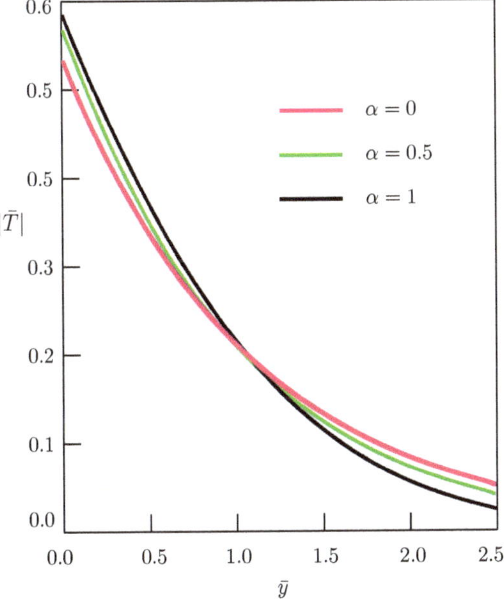

Figure 3. Dependence of the solution on the spatial coordinate y. Typical curves for $0 \leq \alpha \leq 1$; $\kappa = 1$.

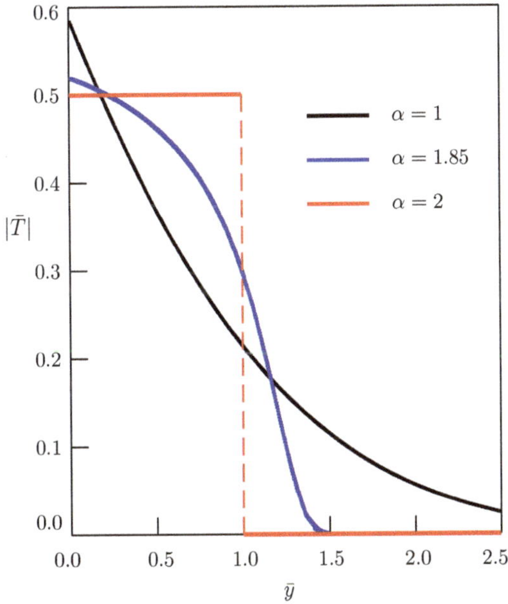

Figure 4. Dependence of the solution on the spatial coordinate y. Typical curves for $1 \leq \alpha \leq 2$; $\kappa = 1$.

4. Conclusions

We have examined the time-fractional heat conduction equation with the Caputo fractional derivative of the order $0 < \alpha \leq 2$ in an infinite plane with two external half-infinite cracks whose surfaces are exposed to the heat flux loading. The solution has been obtained using the integral transform technique. Several special cases corresponding to the integer values of the order of the time-derivative have been considered.

Author Contributions: Both authors have equally contributed to this work. Both authors read and approved the final manuscript.

Funding: This research received no external funding.

Conflicts of Interest: The authors declare no conflict of interest.

References

1. Gurtin, M.E.; Pipkin, A.C. A general theory of heat conduction with finite wave speeds. *Arch. Ration. Mech. Anal.* **1968**, *31*, 113–126. [CrossRef]

2. Nigmatullin, R.R. On the theory of relaxation for systems with "remnant" memory. *Phys. Status Solidi B* **1984**, *124*, 389–393. [CrossRef]

3. Povstenko, Y. Theory of thermoelasticity based on the space-time-fractional heat conduction equation. *Phys. Scr.* **2009**, *136*, 014017. [CrossRef]

4. Povstenko, Y. Fractional Cattaneo-type equations and generalized thermoelasticity. *J. Therm. Stress.* **2011**, *34*, 97–114. [CrossRef]

5. Nigmatullin, R.R. To the theoretical explanation of the "universal response". *Phys. Status Solidi B* **1984**, *123*, 739–745. [CrossRef]

6. Green, A.E.; Naghdi, P.M. Thermoelasticity without energy dissipation. *J. Elast.* **1993**, *31*, 189–208. [CrossRef]

7. Cattaneo, C. Sulla conduzione del calore. *Atti del Seminario Matematico e Fisico dell' Universita di Modena* **1948**, *3*, 83–101.

8. Chandrasekharaiah, D.S. Thermoelasticity with second sound: A review. *Appl. Mech. Rev.* **1986**, *39*, 355–376. [CrossRef]

9. Povstenko, Y. Theories of thermoelasticity based on space-time-fractional Cattaneo-type equations. In Proceedings of the 4th IFAC Workshop on Fractional Differentiation and Its Applications, Badajoz, Spain, 18–26 October 2010.

10. Povstenko, Y. Theories of thermal stresses based on space-time-fractional telegraph equations. *Comput. Math. Appl.* **2012**, *64*, 3321–3328. [CrossRef]

11. Povstenko, Y. Fractional heat conduction equation and associated thermal stresses. *J. Therm. Stress.* **2005**, *28*, 83–102. [CrossRef]

12. Povstenko, Y. Thermoelasticity based on fractional heat conduction equation. In Proceedings of the 6th International Congress on Thermal Stresses, Vienna, Austria, 26–29 May 2005; Volume 2, pp. 501–504.

13. Povstenko, Y. Thermoelasticity which uses fractional heat conduction equation. *J. Math. Sci.* **2009**, *162*, 296–305. [CrossRef]

14. Povstenko, Y. Non-axisymmetric solutions to time-fractional diffusion-wave equation in an infinite cylinder. *Fract. Calc. Appl. Anal.* **2011**, *14*, 418–435. doi:10.2478/s13540-011-0026-4. [CrossRef]

15. Podlubny, I. *Fractional Differential Equations*; Academic Press: San Diego, CA, USA, 1999.

16. Kilbas, A.A.; Srivastava, H.M.; Trujillo, J.J. *Theory and Applications of Fractional Differential Equations*; Elsevier: Amsterdam, The Netherlands, 2006.

17. Magin, R.L. *Fractional Calculus in Bioengineering*; Begell House Publishers, Inc.: Redding, CA, USA, 2006.

18. Mainardi, F. *Fractional Calculus and Waves in Linear Viscoelasticity: An Introduction to Mathematical Models*; Imperial College Press: London, UK, 2010.

19. Tarasov, V.E. *Fractional Dynamics: Applications of Fractional Calculus to Dynamics of Particles, Fields and Media*; Springer: Berlin/Heidelberg, Germany, 2010.

20. Tenreiro Machado, J.; Kiryakova, V.; Mainardi, F. Recent history of fractional calculus. *Commun. Nonlinear Sci. Numer. Simul.* **2011**, *16*, 1140–1153. [CrossRef]

21. Tenreiro Machado, J. And I say to myself: "What a fractional world!" *Fract. Calc. Appl. Anal.* **2011**, *14*, 635–654. doi:10.2478/s13540-011-0037-1 [CrossRef]

22. Datsko, B.; Luchko, Y.; Gafiychuk, V. Pattern formation in fractional reaction-diffusion systems with multiple homogeneous states. *Int. J. Bifurc. Chaos* **2012**, *22*, 1250087. [CrossRef]

23. Uchaikin, V.V. *Fractional Derivatives for Physicists and Engineers*; Springer: Berlin, Germany, 2013.

24. Atanacković, T.M.; Pilipović, S.; Stanković, B.; Zorica, D. *Fractional Calculus with Applications in Mechanics: Vibrations and Diffusion Processes*; John Wiley & Sons: Hoboken, NJ, USA, 2014.

25. Herrmann, R. *Fractional Calculus: An Introduction for Physicists*, 2nd ed.; World Scientific: Singapore, 2014.

26. Povstenko, Y. *Fractional Thermoelasticity*; Springer: New York, NY, USA, 2015.

27. Datsko, B.; Gafiychuk, V.; Podlubny, I. Solitary travelling auto-waves in fractional reaction–diffusion systems. *Commun. Nonlinear Sci. Numer. Simul.* **2015**, *23*, 378–387. [CrossRef]

28. Wyss, W. The fractional diffusion equation. *J. Math. Phys.* **1986**, *27*, 2782–2785. [CrossRef]

29. Mainardi, F. The fundamental solutions for the fractional diffusion-wave equation. *Appl. Math. Lett.* **1996**, *9*, 23–28. [CrossRef]

30. Mainardi, F. Fractional relaxation-oscillation and fractional diffusion-wave phenomena. *Chaos Solitons Fractals* **1996**, *7*, 1461–1477. [CrossRef]

31. Povstenko, Y. *Linear Fractional Diffusion-Wave Equation for Scientists and Engineers*; Birkhäuser: New York, NY, USA, 2015.

32. Olesiak, Z.; Sneddon, I.N. The distribution of thermal stress in an infinite elastic solid containing a penny-shaped crack. *Arch. Ration. Mech. Anal.* **1959**, *4*, 238–254. [CrossRef]

33. Sih, G.C. On the singular character of thermal stresses near a crack tip. *J. Appl. Mech.* **1962**, *29*, 587–589. [CrossRef]

34. Sekine, H. Thermal stress singularities at tips of a crack in a semi-infinite medium under uniform heat flow. *Eng. Fract. Mech.* **1975**, *7*, 713–729. [CrossRef]

35. Kit, G.S.; Krivtsun, M.G. *Plane Problems of Thermoelasticity for Bodies with Cracks*; Naukova Dumka: Kiev, Ukraine, 1983. (In Russian)

36. Nied, H.F. Thermal shock fracture in an edge-cracked plate. *J. Therm. Stress.* **1983**, *6*, 217–229. [CrossRef]

37. Nied, H.F. Thermal shock in an edge-cracked plate subjected to uniform surface heating. *Eng. Fract. Mech.* **1987**, *26*, 239–246. [CrossRef]

38. Kokini, K. Thermal shock of a cracked strip: Effect of temperature-dependent material properties. *Eng. Fract. Mech.* **1986**, *25*, 167–176. [CrossRef]

39. Bahr, H.-A.; Balke, H.; Kuna, M.; Liesk, H. Fracture analysis of a single edge cracked strip under thermal shock. *Theor. Appl. Fract. Mech.* **1987**, *8*, 33–39. [CrossRef]

40. Kerr, G.; Melrose, G.; Tweed, J. The disturbance of a uniform heat flow by a line crack in an infinite anisotropic thermoelastic solid. *Int. J. Eng. Sci.* **1992**, *30*, 1301–1313. [CrossRef]

41. Kit, G.S.; Poberezhnyi, O.V. *Nonstationary Processes in Bodies with Cracklike Defects*; Naukova Dumka: Kiev, Ukraine, 1992. (In Russian)

42. Lam, K.Y.; Tay, T.E.; Yuan, W.G. Stress intensity factors of cracks in finite plates subjected to thermal loads. *Eng. Fract. Mech.* **1992**, *43*, 641–650. [CrossRef]

43. Povstenko, Y.; Kyrylych, T. Fractional thermoelasticity problem for a plane with a line crack under heat flux loading. *J. Therm. Stress.* **2018**, *41*, 1313–1328. [CrossRef]

44. Prudnikov, A.P.; Brychkov, Y.A.; Marichev, O.I. *Integrals and Series: Elementary Functions*; Gordon and Breach Science Publishers: Amsterdam, The Netherlands, 1986.

45. Prudnikov, A.P.; Brychkov, Y.A.; Marichev, O.I. *Integrals and Series: Special Functions*; Gordon and Breach Science Publishers: Amsterdam, The Netherlands, 1986.

46. Abramowitz, M.; Stegun, I.A. (Eds.) *Handbook of Mathematical Functions with Formulas, Graphs and Mathematical Tables*; Dover: New York, NY, USA, 1972.

Article

Time-Fractional Heat Conduction in Two Joint Half-Planes

Yuriy Povstenko [1],* and Joanna Klekot [2]

[1]   Institute of Mathematics and Computer Science, Faculty of Mathematical and Natural Sciences,
      Jan Dlugosz University in Czestochowa, Armii Krajowej 13/15, 42-200 Czestochowa, Poland
[2]   Institute of Mathematics, Faculty of Mechanical Engineering and Computer Science,
      Czestochowa University of Technology, Armii Krajowej 21, 42-200 Czestochowa, Poland;
      joanna.klekot@im.pcz.pl
*   Correspondence: j.povstenko@ajd.czest.pl; Tel.: +48-343-612-269

Received: 4 June 2019; Accepted: 14 June 2019; Published: 16 June 2019

Abstract: The heat conduction equations with Caputo fractional derivative are considered in two joint half-planes under the conditions of perfect thermal contact. The fundamental solution to the Cauchy problem as well as the fundamental solution to the source problem are examined. The Fourier and Laplace transforms are employed. The Fourier transforms are inverted analytically, whereas the Laplace transform is inverted numerically using the Gaver–Stehfest method. We give a graphical representation of the numerical results.

Keywords: fractional calculus; non-Fourier heat conduction; Caputo derivative; Fourier transform; Laplace transform

---

1. Introduction

The conventional theory of heat conduction starts from the Fourier law and, coupled with the energy conservation law, implies the conventional parabolic heat conduction equation. In bodies exhibiting complex features, the usual Fourier law and heat conduction equation are not precise enough, and nonstandard theories, in which these equations are substituted by extended (nonlocal) equations, are elaborated.

Time nonlocal equations give an account of memory. Gurtin and Pipkin [1] suggested the general relation between the heat flux vector **q** and the temperature gradient:

$$\mathbf{q} = -k \int_0^\infty K(\tau)\,\mathrm{grad}\,T(t-\tau)\,\mathrm{d}\tau. \tag{1}$$

In Equation (1), k can be treated as the generalized thermal conductivity, T denotes temperature, t stands for time, and $K(\tau)$ is the weight function.

Nigmatullin [2,3] proposed a similar version of Equation (1):

$$\mathbf{q} = -k \int_0^t K(t-\tau)\,\mathrm{grad}\,T(\tau)\,\mathrm{d}\tau, \tag{2}$$

and obtained for temperature the equation with memory:

$$\frac{\partial T}{\partial t} = a \int_0^t K(t-\tau)\,\Delta T(\tau)\,\mathrm{d}\tau. \tag{3}$$

In Equation (3), a can be treated as the generalized thermal diffusivity coefficient.

Different choices of the weight function $K(t - \tau)$ ("full sclerosis", "short-tail" memory, "middle-tail" memory, "long-tail" memory, and "full memory") were analized in [4–8].

Though different time nonlocal generalizations of the Fourier law have a long history, they still attract the attention of researchers (see, for example, Reference [9], and the extensive discussion in [10]).

Fractional calculus (the theory of integrals and derivatives of non-integer order) has a large number of uses in engineering, geophysics, physics, geology, chemistry, rheology, biology, finance, and medicine (see [11–21], among many others).

Equation (2), written with the "long-tail" power kernel [4–8], can be represented in terms of fractional calculus:

$$\mathbf{q}(t) = -kD_{RL}^{1-\alpha}\text{grad}\, T(t), \qquad 0 < \alpha \leq 1, \tag{4}$$

$$\mathbf{q}(t) = -kI^{\alpha-1}\text{grad}\, T(t), \qquad 1 < \alpha \leq 2. \tag{5}$$

In Equations (4) and (5), $I^\alpha f(t)$ and $D_{RL}^\alpha f(t)$ are the Riemann–Liouville fractional integral and derivative of the order α [22–24]:

$$I^\alpha f(t) = \frac{1}{\Gamma(\alpha)} \int_0^t (t - \tau)^{\alpha-1} f(\tau)\, d\tau, \qquad \alpha > 0, \tag{6}$$

$$D_{RL}^\alpha f(t) = \frac{d^m}{dt^m}\left[\frac{1}{\Gamma(m - \alpha)} \int_0^t (t - \tau)^{m-\alpha-1} f(\tau)\, d\tau\right], \qquad m - 1 < \alpha < m. \tag{7}$$

In Equations (6) and (7), $\Gamma(\alpha)$ denotes the Gamma function.

It should be emphasized that in fractional calculus, there is no sharp boundary between integrals and derivatives. This is why some authors [11,22] do not use a separate notation for the fractional integral $I^\alpha f(t)$, which is designated as $D_{RL}^{-\alpha} f(t)$ with $\alpha > 0$. Making use of this notation, Equations (4) and (5) can be depicted as one time-nonlocal relation:

$$\mathbf{q}(t) = -kD_{RL}^{1-\alpha}\text{grad}\, T(t), \qquad 0 < \alpha \leq 2. \tag{8}$$

Coupled with the law of conservation of energy, the constitutive relation (8) implies the time fractional heat conduction equation:

$$\frac{\partial^\alpha T}{\partial t^\alpha} = a\,\Delta T, \qquad 0 < \alpha \leq 2, \tag{9}$$

with the Caputo fractional derivative:

$$\frac{d^\alpha f(t)}{dt^\alpha} = \frac{1}{\Gamma(m - \alpha)} \int_0^t (t - \tau)^{m-\alpha-1} \frac{d^m f(\tau)}{d\tau^m}\, d\tau, \qquad m - 1 < \alpha < m. \tag{10}$$

The interested reader can find the details of obtaining the time fractional heat conduction Equation (9) from the constitutive relation (8) in [25].

Equations (4) and (9) for $0 < \alpha < 1$ describe the so-called subdiffusion (slow diffusion), which is characterized by the mean-squared displacement $\langle x^2 \rangle \sim t^\alpha$.

Recall the rules of the Laplace transform of fractional integrals and derivatives [22–24]:

$$\mathcal{L}\left\{I^\alpha f(t)\right\} = \frac{1}{s^\alpha}\, f^*(s), \tag{11}$$

$$\mathcal{L}\left\{D_{RL}^\alpha f(t)\right\} = s^\alpha f^*(s) - \sum_{k=0}^{m-1} D^k I^{m-\alpha} f(0^+) s^{m-1-k}, \qquad m - 1 < \alpha < m, \tag{12}$$

$$\mathcal{L}\left\{\frac{d^\alpha f(t)}{dt^\alpha}\right\} = s^\alpha f^*(s) - \sum_{k=0}^{m-1} f^{(k)}(0^+) s^{\alpha-1-k}, \qquad m-1 < \alpha < m. \tag{13}$$

In Equations (11)–(13), the asterisk marks the transform, and s stands for the Laplace transform variable.

A variety of boundary conditions for the time fractional heat conduction equation were analyzed in [8,20,26,27]. If the surfaces of two bodies are in perfect thermal contact, the temperatures at their joint boundary S and the heat fluxes through this boundary are the same for both bodies, and we arrive at the following boundary conditions:

$$T_1\big|_S = T_2\big|_{S'} \tag{14}$$

$$k_1 D_{RL}^{1-\alpha}\frac{\partial T_1}{\partial n}\bigg|_S = k_2 D_{RL}^{1-\beta}\frac{\partial T_2}{\partial n}\bigg|_S, \qquad 0 < \alpha \le 2, \qquad 0 < \beta \le 2, \tag{15}$$

where subscripts 1 and 2 mark bodies 1 and 2, respectively, and n is the common normal at the joint surface. In Equation (15), it is assumed that in the general case, the heat conduction in one of the bodies is described by the equation with the Caputo derivative of the order α, whereas in the second body, the process is governed by the equation with the Caputo derivative of the order β.

In previous publications [27–29], the problems of fractional heat conduction in composed bodies were investigated for one spatial coordinate. In the present paper, heat conduction with the Caputo fractional derivative is considered in two joint half-planes under the conditions of perfect thermal contact in the case of two spatial coordinates. The fundamental solutions to the Cauchy and source problems are studied. The Fourier and Laplace transforms are employed. The Fourier transforms are inverted analytically, and the Laplace transform is inverted numerically, employing the Gaver–Stehfest method [30–33]. The interested reader is also referred to the recent paper [34] on numerical inversion of the Laplace transform for solving fractional deifferenetial equations, where additional references can be found.

2. The Fundamental Solution to the Cauchy Problem

Consider the time fractional heat conduction equations in two half-planes:

$$\frac{\partial^\alpha T_1(x,y,t)}{\partial t^\alpha} = a_1\left[\frac{\partial^2 T_1(x,y,t)}{\partial x^2} + \frac{\partial^2 T_1(x,y,t)}{\partial y^2}\right], \tag{16}$$

$$0 < x < \infty, \quad -\infty < y < \infty, \quad 0 < t < \infty, \quad 0 < \alpha \le 2,$$

$$\frac{\partial^\beta T_2(x,y,t)}{\partial t^\beta} = a_2\left[\frac{\partial^2 T_2(x,y,t)}{\partial x^2} + \frac{\partial^2 T_2(x,y,t)}{\partial y^2}\right], \tag{17}$$

$$-\infty < x < 0, \quad -\infty < y < \infty, \quad 0 < t < \infty, \quad 0 < \beta \le 2.$$

We assume the initial conditions:

$$t = 0: \quad T_1(x,y,t) = p_0\,\delta(x-1)\,\delta(y), \quad 0 < x < \infty, \quad -\infty < y < \infty, \quad 0 < \alpha \le 2, \tag{18}$$

$$t = 0: \quad \frac{\partial T_1(x,y,t)}{\partial t} = 0, \quad 0 < x < \infty, \quad -\infty < y < \infty, \quad 1 < \alpha \le 2, \tag{19}$$

$$t = 0: \quad T_2(x,y,t) = 0, \quad -\infty < x < 0, \quad -\infty < y < \infty, \quad 0 < \beta \le 2, \tag{20}$$

$$t = 0: \quad \frac{\partial T_2(x,y,t)}{\partial t} = 0, \quad -\infty < x < 0, \quad -\infty < y < \infty, \quad 1 < \beta \le 2, \tag{21}$$

and the conditions of perfect thermal contact:

$$x = 0: \quad T_1(x,y,t) = T_2(x,y,t), \quad -\infty < y < \infty, \quad 0 < t < \infty, \quad 0 < \alpha \le 2, \quad 0 < \beta \le 2, \quad (22)$$

$$x = 0: \quad k_1 D_{RL}^{1-\alpha} \frac{\partial T_1(x,y,t)}{\partial x} = k_2 D_{RL}^{1-\beta} \frac{\partial T_2(x,y,t)}{\partial x},$$

$$-\infty < y < \infty, \quad 0 < t < \infty, \quad 0 < \alpha \le 2, \quad 0 < \beta \le 2. \quad (23)$$

The temperatures should also satisfy the zero conditions at infinity:

$$\lim_{x \to \infty} T_1(x,y,t) = 0, \qquad \lim_{y \to \pm\infty} T_1(x,y,t) = 0, \quad (24)$$

$$\lim_{x \to -\infty} T_2(x,y,t) = 0, \qquad \lim_{y \to \pm\infty} T_2(x,y,t) = 0, \quad (25)$$

$$\lim_{t \to \infty} T_1(x,y,t) = 0, \qquad \lim_{t \to \infty} T_2(x,y,t) = 0. \quad (26)$$

We introduce the unknown function:

$$\varphi(y,t) = k_1 D_{RL}^{1-\alpha} \left. \frac{\partial T_1(x,y,t)}{\partial x} \right|_{x=0} = k_2 D_{RL}^{1-\beta} \left. \frac{\partial T_2(x,y,t)}{\partial x} \right|_{x=0}. \quad (27)$$

In the subsequent text, we employ the following cos-Fourier transforms with respect to the coordinate x: for $x > 0$:

$$\mathcal{F}_c\{f(x)\} = \tilde{f}(\xi) = \int_0^\infty f(x) \cos(x\xi)\, dx, \quad (28)$$

$$\mathcal{F}_c^{-1}\{\tilde{f}(\xi)\} = f(x) = \frac{2}{\pi} \int_0^\infty \tilde{f}(\xi) \cos(x\xi)\, d\xi, \quad (29)$$

$$\mathcal{F}_c\left\{ \frac{d^2 f(x)}{dx^2} \right\} = -\xi^2 \tilde{f}(\xi) - \left. \frac{df(x)}{dx} \right|_{x=0^+}, \quad (30)$$

and for $x < 0$:

$$\mathcal{F}_c\{f(x)\} = \tilde{f}(\xi) = \int_{-\infty}^0 f(x) \cos(x\xi)\, dx, \quad (31)$$

$$\mathcal{F}_c^{-1}\{\tilde{f}(\xi)\} = f(x) = \frac{2}{\pi} \int_{-\infty}^0 \tilde{f}(\xi) \cos(x\xi)\, d\xi, \quad (32)$$

$$\mathcal{F}_c\left\{ \frac{d^2 f(x)}{dx^2} \right\} = -\xi^2 \tilde{f}(\xi) + \left. \frac{df(x)}{dx} \right|_{x=0^-}. \quad (33)$$

Applying to the initial boundary value problem (16)–(27) the Laplace transform with respect to time t, the exponential Fourier transform with respect to the coordinate y, and the foregoing cos-Fourier transforms with respect to the coordinate x, in the transform domain we get is:

$$\tilde{\tilde{T}}_1^*(\xi,\eta,s) = \frac{p_0}{\sqrt{2\pi}} \cos(l\xi) \frac{s^{\alpha-1}}{s^\alpha + a_1(\xi^2 + \eta^2)} - \frac{a_1}{k_1} \tilde{\varphi}^*(\eta,s) \frac{s^{\alpha-1}}{s^\alpha + a_1(\xi^2 + \eta^2)}, \quad (34)$$

$$\tilde{\tilde{T}}_2^*(\xi,\eta,s) = \frac{a_2}{k_2} \tilde{\varphi}^*(\eta,s) \frac{s^{\beta-1}}{s^\beta + a_2(\xi^2 + \eta^2)}. \quad (35)$$

All the Fourier transforms are marked by the tilde, ξ denotes the cos-Fourier transform variable, and η stands for the exponential Fourier transform variable.

Inversion of the cos-Fourier transforms with taking into account the integral below [35]:

$$\int_0^\infty \frac{1}{\xi^2 + c^2} \cos(x\xi)\, d\xi = \frac{\pi}{2c} \exp(-c|x|), \quad c > 0, \tag{36}$$

gives:

$$\widetilde{T}_1^*(x,\eta,s) = \frac{p_0}{\sqrt{2\pi}} \frac{s^{\alpha-1}}{2a_1} \frac{1}{\sqrt{\eta^2 + s^\alpha/a_1}} \left\{ \exp\left[-\sqrt{\eta^2 + \frac{s^\alpha}{a_1}}(x+l)\right] + \exp\left[-\sqrt{\eta^2 + \frac{s^\alpha}{a_1}}|x-l|\right] \right\}$$
$$- \frac{s^{\alpha-1}}{k_1} \widetilde{\varphi}^*(\eta,s) \frac{1}{\sqrt{\eta^2 + s^\alpha/a_1}} \exp\left(-\sqrt{\eta^2 + \frac{s^\alpha}{a_1}}x\right), \quad x > 0, \tag{37}$$

$$\widetilde{T}_2^*(x,\eta,s) = \frac{s^{\beta-1}}{k_2} \widetilde{\varphi}^*(\eta,s) \frac{1}{\sqrt{\eta^2 + s^\beta/a_2}} \exp\left(-\sqrt{\eta^2 + \frac{s^\beta}{a_2}}|x|\right), \quad x < 0. \tag{38}$$

The perfect thermal contact boundary condition (19) written in the transform domain:

$$\widetilde{T}_1^*(0,\eta,s) = \widetilde{T}_2^*(0,\eta,s) \tag{39}$$

makes it possible to determine the function: $\widetilde{\varphi}^*(\eta,s)$:

$$\widetilde{\varphi}^*(\eta,s) = \frac{p_0 k_1 k_2}{\sqrt{2\pi}\, a_1} \frac{\sqrt{\eta^2 + s^\beta/a_2}}{s^{\beta-\alpha} k_1 \sqrt{\eta^2 + s^\alpha/a_1} + k_2\sqrt{\eta^2 + s^\beta/a_2}} \exp\left(-\sqrt{\eta^2 + \frac{s^\alpha}{a_1}}\, l\right), \tag{40}$$

and obtain the expressions for temperatures:

$$\widetilde{T}_1^*(x,\eta,s) = \frac{p_0}{\sqrt{2\pi}a_1} \frac{s^{\alpha-1}}{\sqrt{\eta^2 + s^\alpha/a_1}} \left\{ \frac{1}{2}\exp\left[-\sqrt{\eta^2 + \frac{s^\alpha}{a_1}}(x+l)\right] \right.$$
$$+ \frac{1}{2}\exp\left[-\sqrt{\eta^2 + \frac{s^\alpha}{a_1}}|x-l|\right]$$
$$\left. - \frac{k_2\sqrt{\eta^2 + s^\beta/a_2}}{s^{\beta-\alpha}k_1\sqrt{\eta^2 + s^\alpha/a_1} + k_2\sqrt{\eta^2 + s^\beta/a_2}} \exp\left[-\sqrt{\eta^2 + \frac{s^\alpha}{a_1}}(x+l)\right] \right\}, \quad x > 0, \tag{41}$$

$$\widetilde{T}_2^*(x,\eta,s) = \frac{p_0}{\sqrt{2\pi}\, a_1} \frac{k_1 s^{\beta-1}}{s^{\beta-\alpha}k_1\sqrt{\eta^2 + s^\alpha/a_1} + k_2\sqrt{\eta^2 + s^\beta/a_2}}$$
$$\times \exp\left(-\sqrt{\eta^2 + \frac{s^\beta}{a_2}}|x| - \sqrt{\eta^2 + \frac{s^\alpha}{a_1}}\, l\right), \quad x < 0. \tag{42}$$

The inverse transforms give:

$$T_1(x,y,t) = \frac{p_0}{2\pi a_1} \mathcal{L}^{-1}\left\langle \int_{-\infty}^\infty \frac{s^{\alpha-1}}{\sqrt{\eta^2 + s^\alpha/a_1}} \left\{ \frac{1}{2}\exp\left[-\sqrt{\eta^2 + \frac{s^\alpha}{a_1}}(x+l)\right] \right.\right.$$
$$+ \frac{1}{2}\exp\left[-\sqrt{\eta^2 + \frac{s^\alpha}{a_1}}|x-l|\right]$$
$$\left.\left. - \frac{k_2\sqrt{\eta^2 + s^\beta/a_2}}{s^{\beta-\alpha}k_1\sqrt{\eta^2 + s^\alpha/a_1} + k_2\sqrt{\eta^2 + s^\beta/a_2}} \exp\left[-\sqrt{\eta^2 + \frac{s^\alpha}{a_1}}(x+l)\right] \right\} \cos(y\eta)\, d\eta \right\rangle, \quad x > 0, \tag{43}$$

$$T_2(x,y,t) \;=\; \frac{p_0 k_1}{2\pi a_1}\, \mathcal{L}^{-1}\Bigg\{ \int_{-\infty}^{\infty} \frac{s^{\beta-1}}{s^{\beta-\alpha}k_1\sqrt{\eta^2+s^\alpha/a_1}+k_2\sqrt{\eta^2+s^\beta/a_2}}$$
$$\times\, \exp\!\left(-\sqrt{\eta^2+\frac{s^\beta}{a_2}}\,|x|-\sqrt{\eta^2+\frac{s^\alpha}{a_1}}\,l\right)\cos(y\eta)\,d\eta\Bigg\},\quad x<0. \tag{44}$$

Temperatures in joint half-planes are presented in Figures 1–6 for varied combinations of parameters. Calculations have been carried out with the following nondimensional quantities:

$$\bar{x}=\frac{x}{l},\quad \bar{y}=\frac{y}{l},\quad \kappa=\frac{\sqrt{a_1}\,t_0^{\alpha/2}}{l},\quad \bar{T}=\frac{a_1 t_0^\alpha}{p_0}\,T, \tag{45}$$

where t_0 is the characteristic time of the process. In all the figures, the values $\alpha=\beta=0.5$ and $\kappa=1$ have been taken.

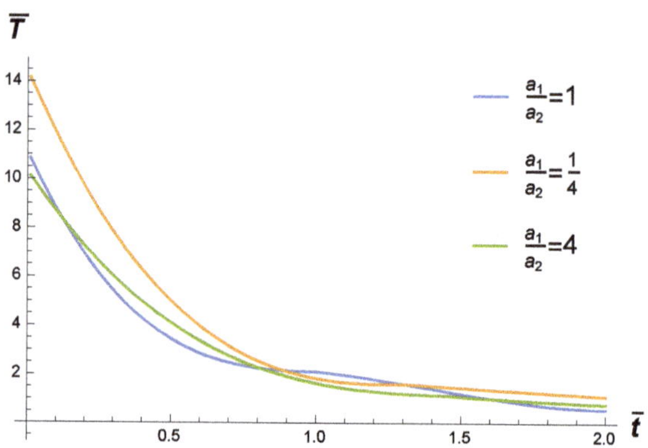

Figure 1. Behavior of the solution (43) and (44) versus time; $\bar{x}=1$, $y=0$, $k_1=k_2$.

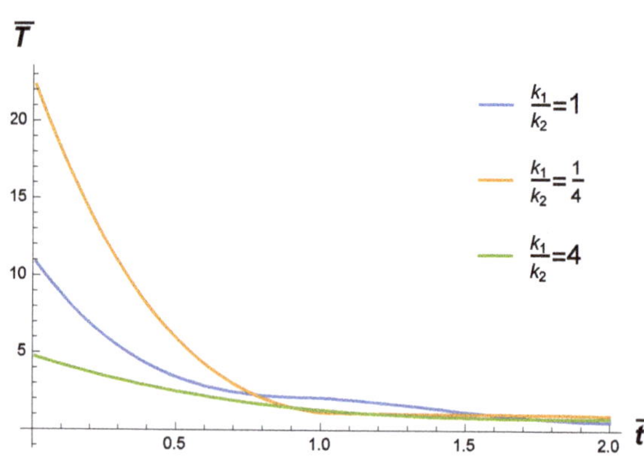

Figure 2. Behavior of the solution (43) and (44) versus time; $\bar{x}=1$, $y=0$, $a_1=a_2$.

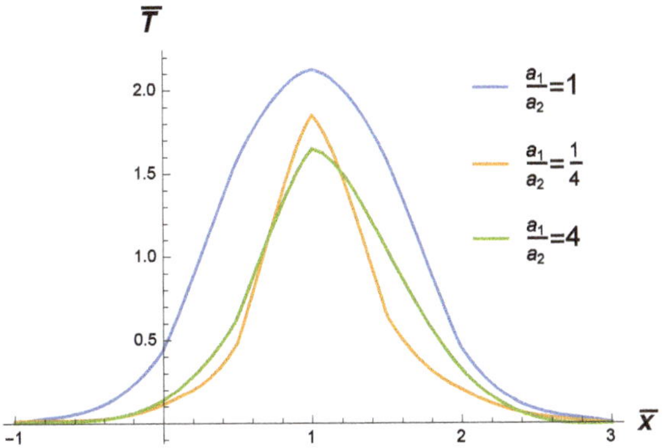

Figure 3. Dependence of the solution (43) and (44) on the coordinate x; $\bar{y} = 0$, $\bar{t} = 1$, $k_1 = k_2$.

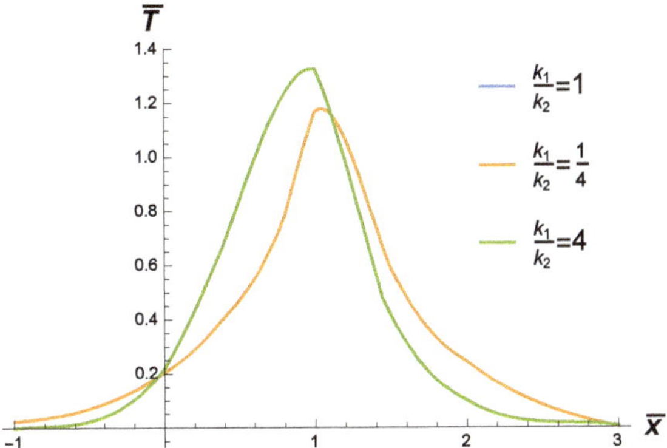

Figure 4. Dependence of the solution (43) and (44) on the coordinate x; $\bar{y} = 0$, $\bar{t} = 1$, $a_1 = a_2$.

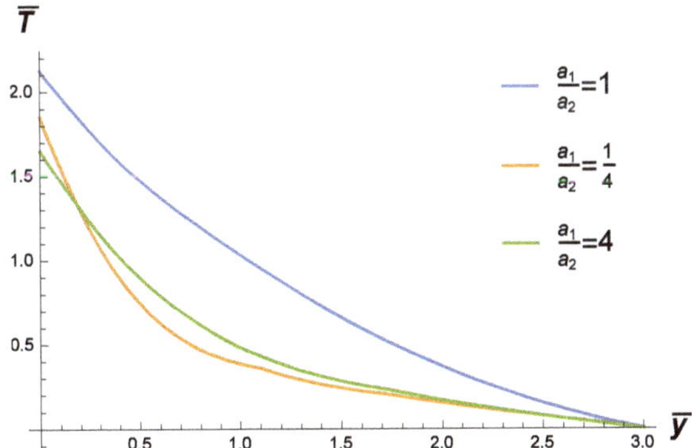

Figure 5. Dependence of the solution (43) and (44) on the coordinate y; $\bar{x} = 1$, $\bar{t} = 1$, $k_1 = k_2$.

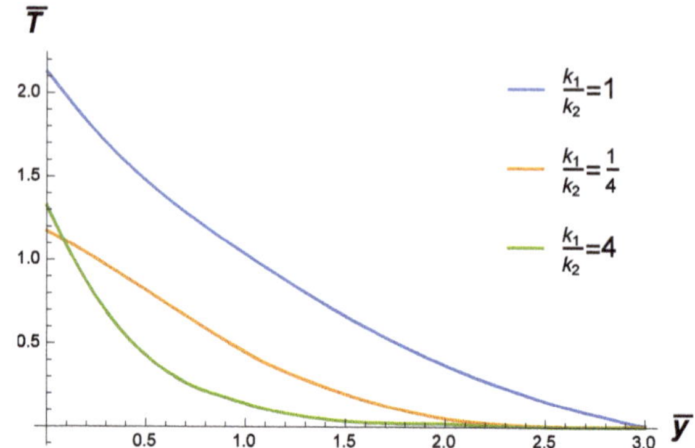

Figure 6. Dependence of the solution (43) and (44) on the coordinate y; $\bar{x} = 1$, $\bar{t} = 1$, $a_1 = a_2$.

3. The Fundamental Solution to the Source Problem

Next, we consider the time fractional heat conduction equation with the source term in the domain $x > 0$:

$$\frac{\partial^\alpha T_1(x,y,t)}{\partial t^\alpha} = a_1 \left[\frac{\partial^2 T_1(x,y,t)}{\partial x^2} + \frac{\partial^2 T_1(x,y,t)}{\partial y^2} \right] + g_0\, \delta\,(x - l)\, \delta\,(y)\, \delta\,(t), \tag{46}$$

$$0 < x < \infty, \quad -\infty < y < \infty, \quad 0 < t < \infty, \quad 0 < \alpha \le 2,$$

and the corresponding equation in the region $x < 0$:

$$\frac{\partial^\beta T_2(x,y,t)}{\partial t^\beta} = a_2 \left[\frac{\partial^2 T_2(x,y,t)}{\partial x^2} + \frac{\partial^2 T_2(x,y,t)}{\partial y^2} \right], \tag{47}$$

$$-\infty < x < 0, \quad -\infty < y < \infty, \quad 0 < t < \infty, \quad 0 < \beta \le 2.$$

We assume the zero initial conditions:

$$t = 0: \quad T_1(x,y,t) = 0, \quad 0 < x < \infty, \quad -\infty < y < \infty, \quad 0 < \alpha \le 2, \tag{48}$$

$$t = 0: \quad \frac{\partial T_1(x,y,t)}{\partial t} = 0, \quad 0 < x < \infty, \quad -\infty < y < \infty, \quad 1 < \alpha \le 2, \tag{49}$$

$$t = 0: \quad T_2(x,y,t) = 0, \quad -\infty < x < 0, \quad -\infty < y < \infty, \quad 0 < \beta \le 2, \tag{50}$$

$$t = 0: \quad \frac{\partial T_2(x,y,t)}{\partial t} = 0, \quad -\infty < x < 0, \quad -\infty < y < \infty, \quad 1 < \beta \le 2, \tag{51}$$

and the conditions of perfect thermal contact at the joint boundary:

$$x = 0: \quad T_1(x,y,t) = T_2(x,y,t), \quad -\infty < y < \infty, \quad 0 < t < \infty, \quad 0 < \alpha \le 2, \quad 0 < \beta \le 2, \tag{52}$$

$$x = 0: \quad k_1 D_{RL}^{1-\alpha} \frac{\partial T_1(x,y,t)}{\partial x} = k_2 D_{RL}^{1-\beta} \frac{\partial T_2(x,y,t)}{\partial x},$$

$$-\infty < y < \infty, \quad 0 < t < \infty, \quad 0 < \alpha \le 2, \quad 0 < \beta \le 2. \tag{53}$$

Using the integral transform technique, we finally obtain:

$$
\begin{aligned}
T_1(x,y,t) = \frac{g_0}{2\pi a_1}\, \mathcal{L}^{-1}\Bigg\langle \int_{-\infty}^{\infty} \frac{1}{\sqrt{\eta^2 + s^\alpha/a_1}} \Bigg\{ \frac{1}{2}\exp\left[-\sqrt{\eta^2 + \frac{s^\alpha}{a_1}}\,(x+l)\right] \\
+ \frac{1}{2}\exp\left[-\sqrt{\eta^2 + \frac{s^\alpha}{a_1}}\,|x-l|\right] \\
- \frac{k_2\sqrt{\eta^2 + s^\beta/a_2}}{s^{\beta-\alpha}k_1\sqrt{\eta^2 + s^\alpha/a_1} + k_2\sqrt{\eta^2 + s^\beta/a_2}}\exp\left[-\sqrt{\eta^2 + \frac{s^\alpha}{a_1}}\,(x+l)\right]\Bigg\} \cos(y\eta)\,\mathrm{d}\eta \Bigg\rangle, \quad x > 0,
\end{aligned}
\tag{54}
$$

$$
\begin{aligned}
T_2(x,y,t) = \frac{g_0 k_1}{2\pi a_1}\, \mathcal{L}^{-1}\Bigg\{ \int_{-\infty}^{\infty} \frac{s^{\beta-\alpha}}{s^{\beta-\alpha}k_1\sqrt{\eta^2 + s^\alpha/a_1} + k_2\sqrt{\eta^2 + s^\beta/a_2}} \\
\times \exp\left(-\sqrt{\eta^2 + \frac{s^\beta}{a_2}}\,|x| - \sqrt{\eta^2 + \frac{s^\alpha}{a_1}}\,l\right)\cos(y\eta)\,\mathrm{d}\eta \Bigg\}, \quad x < 0.
\end{aligned}
\tag{55}
$$

The numerical results according to Equations (54) and (55) with $\alpha = \beta = 0.5$, $\kappa = 1$ are presented in Figures 7–12. The nondimensional temperature is introduced as:

$$
\bar{T} = \frac{a_1 t_0}{g_0},
\tag{56}
$$

while other nondimensional quantities are the same as in Equation (45).

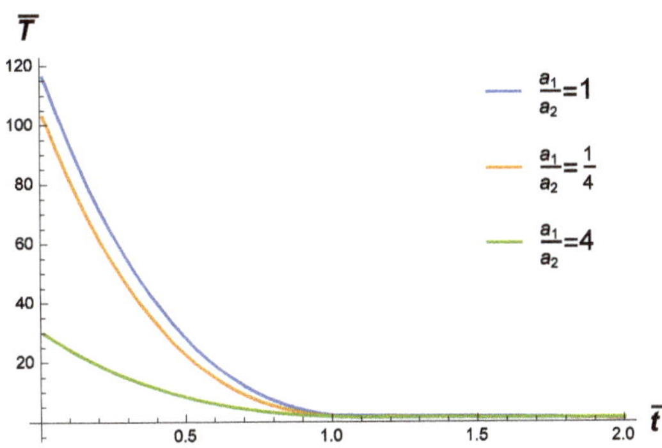

Figure 7. Behavior of the solution (54) and (55) versus time; $\bar{x} = 1$, $y = 0$, $k_1 = k_2$.

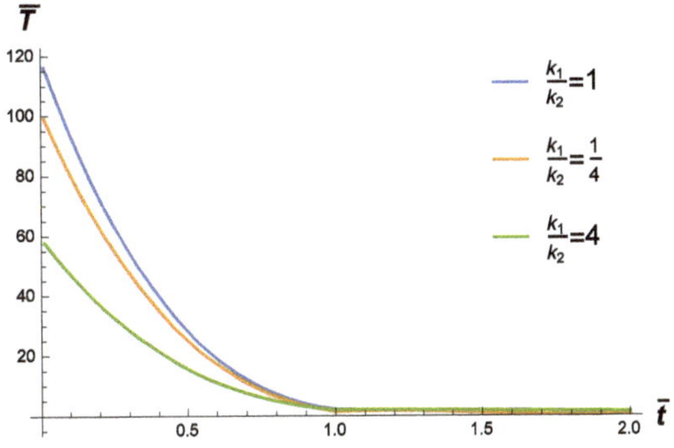

Figure 8. Behavior of the solution (54) and (55) versus time; $\bar{x} = 1$, $y = 0$, $a_1 = a_2$.

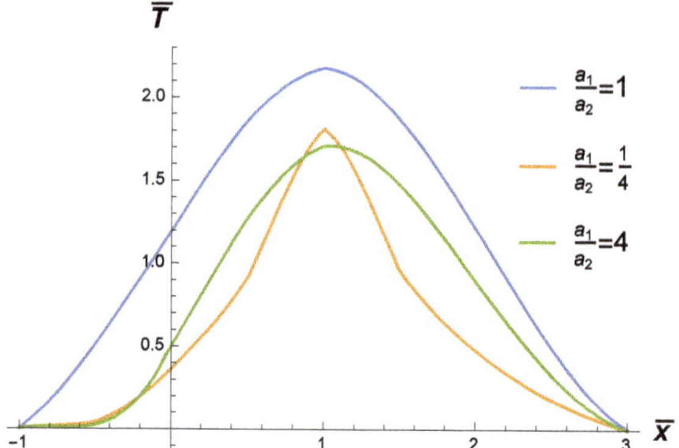

Figure 9. Dependence of the solution (54) and (55) on the coordinate x; $\bar{y} = 0, \bar{t} = 1, k_1 = k_2$.

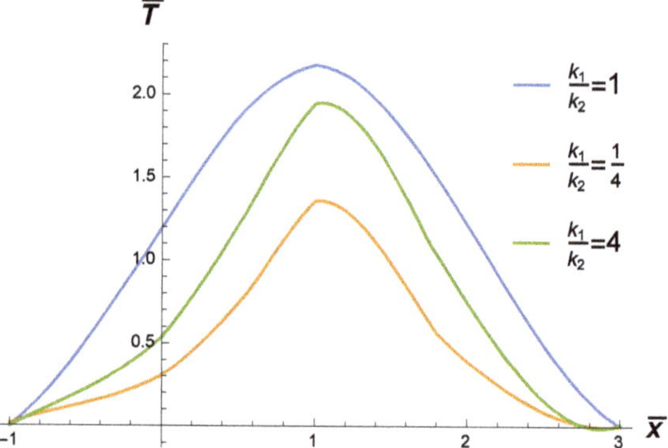

Figure 10. Dependence of the solution (54) and (55) on the coordinate x; $\bar{y} = 0, \bar{t} = 1, a_1 = a_2$.

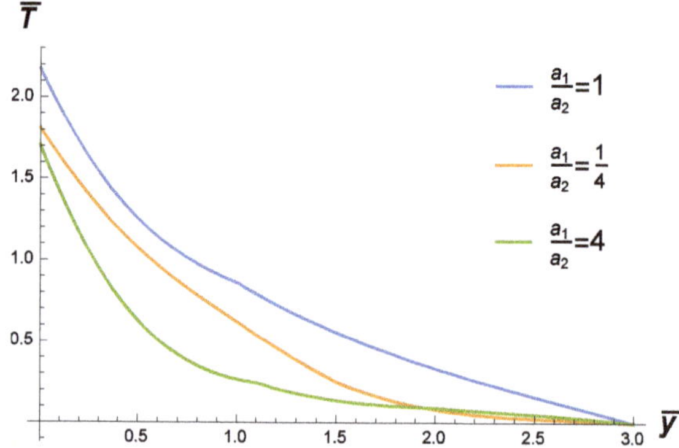

Figure 11. Dependence of the solution (54) and (55) on the coordinate y; $\bar{x} = 1, \bar{t} = 1, k_1 = k_2$.

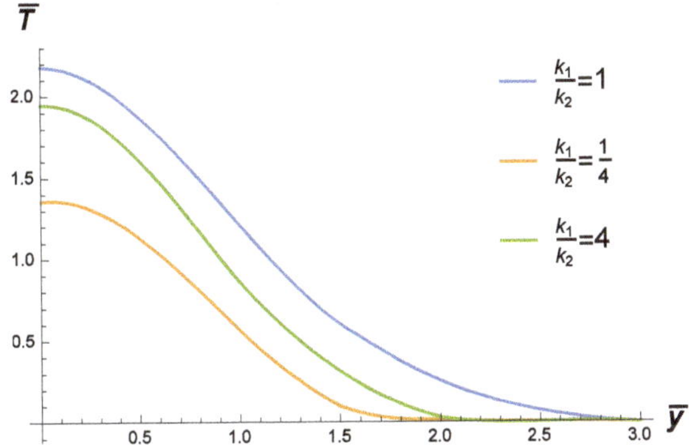

Figure 12. Dependence of the solution (54) and (55) on the coordinate y; $\bar{x}=1$, $\bar{t}=1$, $a_1=a_2$.

4. Conclusions

We studied the heat conduction equations with the Caputo fractional derivatives in two joint half-planes under conditions of perfect thermal contact (the equality of temperatures and heat fluxes at the contact surface). It should be emphasized that due to the constitutive Equation (8), the proper boundary conditions should be stated in terms of the heat fluxes, not in terms of the normal derivatives of temperature alone. Introducing the auxiliary function $\phi(x,t)$ allows us to use the cos-Fourier transforms in two contact regions. The fundamental solutions permit obtaining various solutions to the Cauchy and source problems in the convolution form.

Author Contributions: Both authors have equally contributed to this work. Both authors read and approved the final manuscript.

Funding: This research received no external funding.

Acknowledgments: The authors would like to thank the reviewers for their helpful comments to improve the quality of the paper.

Conflicts of Interest: The authors declare no conflict of interest.

References

1. Gurtin, M.E.; Pipkin, A.C. A general theory of heat conduction with finite wave speeds. *Arch. Ration. Mech. Anal.* **1968**, *31*, 113–126. [CrossRef]
2. Nigmatullin, R.R. To the theoretical explanation of the "universal response". *Phys. Stat. Sol.* **1984**, *123*, 739–745. [CrossRef]
3. Nigmatullin, R.R. On the theory of relaxation for systems with "remnant" memory. *Phys. Stat. Sol.* **1984**, *124*, 389–393. [CrossRef]
4. Povstenko, Y. Fractional heat conduction equation and associated thermal stresses. *J. Therm. Stress.* **2005**, *28*, 83–102. [CrossRef]
5. Povstenko, Y. Thermoelasticity which uses fractional heat conduction equation. *J. Math. Sci.* **2009**, *162*, 296–305. [CrossRef]
6. Povstenko, Y. Theory of thermoelasticity based on the space-time fractional heat conduction equation. *Phys. Scr.* **2009**, *136*, 014017. [CrossRef]
7. Povstenko, Y. Fractional Cattaneo-type equations and generalized thermoelasticity. *J. Therm. Stress.* **2011**, *34*, 97–114. [CrossRef]
8. Povstenko, Y. Fractional thermoelasticity. In *Encyclopedia of Thermal Stresses*; Hetnarski, R.B., Ed.; Springer: New York, NY, USA, 2014; Volume 4, pp. 1778–1787.
9. Rukolaine, S.A. Unphysical effects of the dual-phase-lag model of heat conduction. *Int. J. Heat Mass Transf.* **2014**, *78*, 58–63. [CrossRef]

10. Kovács, R.; Ván, P. Thermodynamical consistency of the dual-phase-lag heat conduction equation. *Contin. Mech. Thermodyn.* **2018**, *30*, 1223–1230. [CrossRef]

11. Magin, R.L. *Fractional Calculus in Bioengineering*; Begell House Publishers, Inc.: Redding, CA, USA, 2006.

12. Sabatier, J., Agrawal, O.P., Tenreiro Machado, J.A. (Eds.) *Advances in Fractional Calculus: Theoretical Developments and Applications in Physics and Engineering*; Springer: Dordrecht, The Netherlands, 2007.

13. Baleanu, D., Güvenç, Z.B., Tenreiro Machado, J.A. (Eds.) *New Trends in Nanotechnology and Fractional Calculus Applications*; Springer: Dordrecht, The Netherlands, 2010.

14. Mainardi, F. *Fractional Calculus and Waves in Linear Viscoelasticity: An Introduction to Mathematical Models*; Imperial College Press: London, UK, 2010.

15. Tarasov, V.E. *Fractional Dynamics: Applications of Fractional Calculus to Dynamics of Particles, Fields and Media*; Springer: Berlin/Heidelberg, Germany, 2010.

16. Datsko, B.; Luchko, Y.; Gafiychuk, V. Pattern formation in fractional reaction-diffusion systems with multiple homogeneous states. *Int. J. Bifurcat. Chaos* **2012**, *22*, 1250087. [CrossRef]

17. Uchaikin, V.V. *Fractional Derivatives for Physicists and Engineers*; Springer: Berlin, Germany, 2013.

18. Atanacković, T.M.; Pilipović, S.; Stanković, B.; Zorica, D. *Fractional Calculus with Applications in Mechanics: Vibrations and Diffusion Processes*; John Wiley & Sons: Hoboken, NJ, USA, 2014.

19. Herrmann, R. *Fractional Calculus: An Introduction for Physicists*, 2nd ed.; World Scientific: Singapore, 2014.

20. Povstenko, Y. *Fractional Thermoelasticity*; Springer: New York, NY, USA, 2015.

21. Datsko, B.; Gafiychuk, V.; Podlubny, I. Solitary travelling auto-waves in fractional reaction–diffusion systems. *Commun. Nonlinear Sci. Numer. Simul.* **2015**, *23*, 378–387. [CrossRef]

22. Podlubny, I. *Fractional Differential Equations*; Academic Press: San Diego, CA, USA, 1999.

23. Kilbas, A.A.; Srivastava, H.M.; Trujillo, J.J. *Theory and Applications of Fractional Differential Equations*; Elsevier: Amsterdam, The Netherlands, 2006.

24. Povstenko, Y. *Linear Fractional Diffusion-Wave Equation for Scientists and Engineers*; Birkhäuser: New York, NY, USA, 2015.

25. Povstenko, Y. Non-axisymmetric solutions to time-fractional diffusion-wave equation in an infinite cylinder. *Fract. Calc. Appl. Anal.* **2011**, *14*, 418–435, doi:10.2478/s13540-011-0026-4. [CrossRef]

26. Povstenko, Y. Neumann boundary-value problems for a time-fractional diffusion-wave equation in a half-plane. *Comput. Math. Appl.* **2012**, *64*, 3183–3192. [CrossRef]

27. Povstenko, Y. Fractional heat conduction in infinite one-dimensional composite medium. *J. Therm. Stress.* **2013**, *36*, 351–363. [CrossRef]

28. Povstenko, Y. Fractional heat conduction in an infinite medium with a spherical inclusion. *Entropy* **2013**, *15*, 4122–4133. [CrossRef]

29. Povstenko, Y. Fundamental solutions to time-fractional heat conduction equations in two joint half-lines. *Cent. Eur. J. Phys.* **2013**, *11*, 1284–1294. [CrossRef]

30. Gaver, D.P. Observing stochastic processes, and approximate transform inversion. *Oper. Res.* **1966**, *14*, 444–459. [CrossRef]

31. Stehfest, H. Algorithm 368 Numerical inversion of Laplace transform. *Commun. ACM* **1970**, *13*, 47–49. [CrossRef]

32. Stehfest, H. Remark on algorithm 368 Numerical inversion of Laplace transform. *Commun. ACM* **1970**, *13*, 624. [CrossRef]

33. Kuznetsov, A. On the convergence of Gaver–Stehfest algorithm. *SIAM J. Numer. Anal.* **2013**, *51*, 2984–2998. [CrossRef]

34. Rani, D.; Mishra, V.; Cattani, C. Numerical inverse Laplace transform for solving a class of fractional differential equations. *Symmetry* **2019**, *11*, 530. [CrossRef]

35. Prudnikov, A.P.; Brychkov, Y.A.; Marichev, O.I. *Integrals and Series, Vol 1: Elementary Functions*; Gordon and Breach Science Publishers: Amsterdam, The Netherlands, 1986.

MDPI

St. Alban-Anlage 66

4052 Basel

Switzerland

Tel. +41 61 683 77 34

Fax +41 61 302 89 18

www.mdpi.com

Symmetry Editorial Office

E-mail: symmetry@mdpi.com

www.mdpi.com/journal/symmetry